Optimization of Membrane Treatment for Direct and Clarified Water Filtration

Optimization of Membrane Treatment for Direct and Clarified Water Filtration

Prepared by:
Samer Adham, Kuang-ping Chiu, and Geno Lehman
MWH
301 North Lake Avenue, Suite 600
Pasadena, California, 91101, USA

Kerry Howe, and Ashish Marwah
Department of Civil Engineering
University of New Mexico
Albuquerque, NM, 87131-1351, USA

Chandra Mysore, and Jeremie Clouet
Veolia Water
Two Sun Court, Suite 220
Norcross, GA, 30092, USA

and

Zdravka Do-Quang, and Olivier Cagnard
Suez Environment
38 Rue du President Wilson
Le Pecq, 78230, France

Jointly sponsored by:
Awwa Research Foundation
6666 West Quincy Avenue
Denver, CO 80235
and
U.S. Environmental Protection Agency
Washington D.C.

Published by
Awwa Research Foundation

DISCLAIMER

This study was jointly funded by the Awwa Research Foundation (AwwaRF) and the U.S. Environmental Protection Agency (USEPA) under Cooperative Agreement No. R829409-01. AwwaRF and USEPA assume no responsibility for the content of the research study reported in this publication or for the opinions or statements of fact expressed in the report. The mention of trade names for commercial products does not represent or imply the approval or endorsement of AwwaRF or USEPA. This report is presented solely for informational purposes.

CONTENTS

TABLES

FIGURES

FOREWORD

The Awwa Research Foundation is a nonprofit corporation that is dedicated to the implementation of a research effort to help utilities respond to regulatory requirements and traditional high-priority concerns of the industry. The research agenda is developed through a process of consultation with subscribers and drinking water professionals. Under the umbrella of a Strategic Research Plan, the Research Advisory Council prioritizes the suggested projects based upon current and future needs, applicability, and past work; the recommendations are forwarded to the Board of trustees for final selection. The foundation also sponsors research projects through the unsolicited proposal process; the Collaborative Research, Research Applications, and Tailored Collaboration programs; and various joint research efforts with organizations such as the U.S. Environmental Protection Agency, the U.S. Bureau of reclamation, and the Association of California Water Agencies.

This publication is a result of one of these sponsored studies, and it is hoped that its findings will be applied in communities throughout the world. The following report serves not only as a means of communicating the results of the water industry's centralized research program but also as a tool to enlist the further support of the nonmember utilities and individuals.

Projects are managed closely from their inception to the final report by the foundation's staff and large cadre of volunteers who willingly contribute their time and expertise. The foundation serves a planning and management function, and awards contracts to other institutions such as water utilities, universities, and engineering firms. The funding for this research effort comes primarily from the Subscription Program, through which water utilities subscribe to the research program and make an annual payment proportionate to the volume of water they deliver and consultants and manufacturers subscribe based on their annual billings. The program offers a cost-effective and fair method for funding research in the public interest.

A broad spectrum of water supply issues is addressed by the foundation's research agenda: resources, treatment and operations, distribution and storage, water quality and analysis, toxicology, economics, and management. The ultimate purpose of the coordinated effort is to assist water suppliers to provide the highest possible quality of water economically and reliably. The true benefits are realized when the results are implemented at the utility level. The foundation's trustees are pleased to offer this publication as a contribution toward that end.

Walter J. Bishop
Chair, Board of Trustees
Awwa Research Foundation

Robert C. Renner, P.E.
Executive Director
Awwa Research Foundation

ACKNOWLEDGMENTS

The authors are indebted to Awwa Research Foundation (AwwaRF) and U.S. Environmental Protection Agency (USEPA) for funding this project. The authors wish to thank Traci Case, the project manager for AwwaRF and the following members of the Project Advisory Committee: Phil Brandhuber, HDR, Denver, Colo., USA; Peter Hillis, United Utilities Service Delivery, Warrington, UK; Randy Houston, San Jose Water Company, San Jose, Calif., USA; Marie Owens, Metropolitan Water District, Salt Lake City, Utah, USA; Dan Williams, USEPA, Cincinnati, Ohio; for their guidance and suggestions.

The authors would like to acknowledge the participating water utilities that provided water for bench scale testing, provided testing site and technical assistance for pilot testing, and provided data for full-scale case studies. The list below included the water utilities that contributed to this project as well as the manufacturers that assisted in this project.

- Ashland Microfiltration Water Treatment Plant, Wis., USA: Craig Nordgren, and Dave Wosepka
- Atlanta Fulton County Water Treatment Plant, Ga., USA: Mike Leonnard
- BexarMet Water District, Texas, USA: Michael Albach
- Memcor Australia PTY. LTD., Australia: Warren Johnson
- North Clackamas County Water Commission Plant, Ore., USA: John Thomas, Dave Davis, and Alan Schacht
- Suez Environment, France: Isabelle Lemoigne, Patrick Sauvade, Véronique Bonnelye
- United Water, Texas, USA: Joe Thaxton
- USFilter, USA: Paul Gallagher, and Aaron Balczewski
- Veolia Water, Australia: Rod Smith
- Veolia Water, France: Jerome Leparc
- Vigneux Water Treatment Plant, France: Jean-Marc Ponté
- Waikato Water Treatment Plant, New Zealand: Derrick Adams, and Shayne Cunis
- Westminster Northhwest Water Treatment Facility, Colo., USA: Tom Settle

The insight and comments from the following members of the Technical Advisory Committee were also appreciated: Mark Clark, University of Illinois, Urbana-Champaign, Ill.; David Eddy, MWH, Buckinghamshire, UK; Tony Fane, University of New South Wales, Australia; Rick Sakaji, California Department of Health Services, Berkeley, Calif., USA; and Rhodes Trussell, Trussell Technologies, Pasadena, Calif., USA.

The authors of this report would also like to acknowledge the following members of the Utility Advisory Committee for their cooperation and participation in this project: Bruce Bartley, NSF International, Ann Arbor, Mich., USA; Michael Buettner, Appleton Water Treatment Plant, Menasha, Wis., USA; Ed Cummings, Zone 7 Water Agency, Livermore, Calif., USA; Derrick Adams, and Shayne Cunis, Watercare Services, New Zealand; Graham James, Scottish Water, UK; Bill Pearce, City of San Diego, Calif., USA; Jim Naismith and Don Roach, San Patricio MWD, Texas, USA.

Finally, the authors would like to acknowledge the work of Kathryn Mallon, and Jude Ground from MWH, Portland, Ore., as well as Karl Gramith, Brian Gallagher, Eric Bruce, Joan Oppenheimer, Darren Giles, Natalie Flores, and Judy Richardson from MWH, Pasadena, Calif.

EXECUTIVE SUMMARY

BACKGROUND

Low pressure microfiltration (MF) and ultrafiltration (UF) membranes have been established as viable drinking water treatment technologies. Early applications of MF/UF relied on cartridge filtration to protect the membranes. As the success of these stand-alone MF and UF membrane installations increased over time, many utilities with a wide variety of source waters became interested in the application of MF/UF for control of disinfection by-product precursors or removal of natural organic matter (NOM), color, taste and odor, hardness, iron, manganese, arsenic, and other organic and inorganic compounds. These source waters, however, are often more challenging for membranes, as various constituents in the raw water, such as high turbidity and/or high organics, can lead to increased membrane fouling. In addition, MF/UF membranes alone do not always provide adequate treatment for many of contaminants mentioned above. Consequently, additional pretreatment processes must be considered to effectively treat the challenging water supplies and maintain the productivity of the membrane process.

Many utilities also began to consider integrating membranes into their existing conventional treatment plants for various reasons: increasing plant capacity, meeting more stringent regulatory requirements, producing consistent effluent quality and upgrading plant infrastructure. Integrating membranes into an existing treatment plant presents many challenges such as identifying the best location for integration with the treatment train and determining the compatibility of the membrane material and system configuration with the chemicals and infrastructure used in the existing treatment plant.

Unfortunately, the successes of membrane installations without pretreatment are not directly transferable to more challenging water supplies and/or system applications. Insufficient information is available about the interaction between raw/clarified water constituents and how water treatment chemicals will affect membrane performance. Thus, understanding the effect of pretreatment processes on MF/UF membranes is critical to the future applications of membranes in water treatment.

PROJECT OBJECTIVE AND APPROACH

The objective of this project was to investigate the interaction of water quality, pretreatment chemicals, pretreatment processes, membrane configurations, and membrane materials on the membrane performance. The findings of this study can be used as a guidance for the drinking water community to determine how to optimize the integration of MF/UF technologies into existing or new conventional treatment plants. This project was implemented in three phases: bench-, pilot-, and full-scale studies.

Bench-Scale Testing

The primary focus of the bench-scale testing was to examine membrane fouling mechanisms by specific fractions of constituents in natural water using several source waters, membrane materials, and membrane configurations. Bench-scale testing was conducted on five different water sources: (1) Medina River water from the BexarMet Water Treatment Plant (BexarMet) in Texas, (2) Chattahoochee River water from the Atlanta Fulton County Water

Treatment Plant (AFCWTP) in Geogia, (3) Seine River water from the Centre International de Recherche sur l'Eau et l' Environnement (CIRSEE) in Le Pecq, France, (4) Clackamas River water from the North Clackamas County Water Commission (NCCWC) plant in Oregon, and (5) Waikato River water from the Waikato Water Treatment Plant in New Zealand. The general raw water quality data of these water sources are listed in the table below (Table ES.1).

Table ES.1
Raw Water Quality Data From Bench-Scale Testing Water Sources

Parameter	Medina River	Chattahoochee River	Seine River	Clackamas River	Waikato River
pH	8.4	6.9	7.2	7.1	7.5
Alkalinity (mg/L as $CaCO_3$)	219	25	215	10	40
Hardness (mg/L as $CaCO_3$)	248	18	236	23	55
TOC (mg/L)	1.36	1.23	3.01	0.89	1.91
UV_{254} (cm^{-1})	0.054	0.031	0.068	0.067	0.069
SUVA (L/mg·m)	3.9	2.5	2.3	7.5	3.6
Silica (mg/L as SiO_2)	15.4	6.40	2.81	19.6	33.7

The constituents in raw water were fractionated by prefiltration to remove materials larger than 1 μm, 100 kilodalton (kDa, equal to a molecular weight of 1000 grams/mole) and 10 kDa. The effect of various pretreatment chemicals, pretreatment processes on each fraction, and the subsequent effect on membrane performance, were then assessed. Membrane performance was determined by measuring flux decline during constant-pressure filtration using bench-scale membrane filtration equipment. Water quality before and after pretreatment was also measured. Scanning electron microscope (SEM) and Fourier transform infrared (FTIR) spectrometry analysis were performed on some membranes to investigate the characteristics of fouling materials.

Pilot-Scale Study

The goal of the pilot-scale testing was to investigate membrane performance using similar pretreatment conditions of the bench-scale study. Pilot testing was conducted at three different sites: (1) the AFCWTP in Georgia with Chattahoochee River water; (2) CIRSEE in Le Pecq, France with Seine River water; and (3) NCCWC plant in Oregon with Clackamas River water. These tests provided the opportunity to compare the performance of different membrane materials (polyvinylidene fluoride, polysulfone, and cellulose acetate), and pretreatment processes (coagulation, flocculation, and sedimentation).

Full-Scale Evaluation

The primary focus of the evaluation of full-scale facilities was to provide insight on operational and design issues that were impacted by membrane pretreatment. Several worldwide membrane filtration plants with pretreatment were investigated in this project: (1) BexarMet Water Treatment Plant, Texas, USA; (2) Vigneux Water Treatment Plant, France; (3) Ashland Water Treatment Plant, Wisconsin, USA; (4) Meredith Water Treatment Plant, Victoria, Australia; (5) Westminster Water Treatment Plant, Colorado, USA; and (6) Waikato Water Treatment Plant, New Zealand. For each plant, the long-term water quality parameters, membrane flux, chemical cleaning, and pretreatment chemical data were analyzed to evaluate the membrane performance.

CONCLUSIONS

The initial project goal was to develop a unified understanding of the interaction of water quality, membrane materials, pretreatment chemicals, and pretreatment processes. However, as the research was being conducted, the project team discovered that a multitude of variables existed that could impact such interactions, making the original project goal nearly unattainable. Thus, it was decided to focus the bench and pilot-scale testing on selected conditions that were relevant to current applications of MF/UF facilities which require pretreatment. The key findings of the bench- and pilot-scale testing are summarized below (Tables ES.2 and ES.3) and should be used by utilities, researchers, and the water industry as a preliminary tool to guide their understanding of the impact of pretreatment on MF/UF processes. Future research studies can be built from this tool in order to develop a more comprehensive understanding to optimize pretreatment of MF/UF applications. These findings have also been supplemented with a comprehensive table summarizing selected literature (Appendix A) and case studies from full-scale plants to further assist in understanding the various interactions that impact pretreatment and membrane performance of past and current MF/UF facilities.

Table ES.2 Results of Bench-Scale Testing

		Medina River	Seine River	Chattahoochee River	Clackamas River	Waikato River
PS	A	– moderate fouling	– high fouling	–	– high fouling	– high fouling
	B	– clarification reduced fouling – alum was better than ferric	– clarification reduced fouling – alum was better than ferric – PAC at 2 mg/L did not improve membrane performance	–	– clarification reduced fouling – ACH was better than alum and ferric – chlorine increased fouling at short term	–
	C	–	–	–	–	–
	D	–	– flat sheet membrane fouled faster than hollow fiber membrane	–	– flat sheet membrane fouled faster than hollow fiber membrane	– flat sheet membrane fouled faster than hollow fiber membrane
	E	–	–	–	–	–
PVdF	A	– moderate fouling	–	– high fouling	– low fouling	– high fouling
	B	– clarification reduced fouling – PACl was better than alum and ferric	–	– clarification reduced fouling	– clarification increased fouling – ACH was better than alum and ferric – chlorine did not improve membrane performance at short term testing	– clarification did not improve membrane performance
	C	–	–	– best membrane performance was observed on clarified water followed by flocculated water, then coagulated water	–	–
	D	–	–	– flat sheet membrane fouled faster than hollow fiber membrane	– flat sheet and hollow fiber membranes fouled at a similar rate	–
	E	–	–	–	– hydrophilic PVdF fouled less than hydrophobic PVdF	–
CA	A	– moderate fouling	–	–	–	–
	B	– clarification reduced fouling – PACl was better than alum and ferric	–	–	–	–
	C	–	–	–	–	–
	D	–	–	–	–	–
	E	–	–	–	–	–

A: impact of water quality B: impact of pretreatment chemicals C: impact of pretreatment processes D: impact of configurations E: impact of membrane materials

Table ES.3 Results of Pilot-Scale Testing

		Medina River	Seine River	Chattahoochee River	Clackamas River	Waikato River
PS	A	–	–	–	–	-
	B	–	–	–	–	
	C	–	–	–	–	
	D	–	–	–	–	-
	E	–	– PS fouled slightly faster than CA	–	–	
PVdF	A	–	–	– moderate fouling	– moderate fouling	-
	B	–	–	– clarification reduced fouling	– clarification at low coagulant dose slightly reduced fouling – clarification at higher coagulant dose increased fouling – ACH was better than alum – chlorine reduced fouling at long term testing – PAC at high dose increased fouling	-
	C	–	–	– best membrane performance was observed on clarified water followed by flocculated water, then coagulated water	– best membrane performance was observed on clarified water followed by flocculated water, then coagulated water	-
	D	–	–	–	–	-
	E	–	–	– hydrophilic PVdF fouled less than hydrophobic PVdF	–	-
CA	A	–	– high fouling	–	–	-
	B	–	– clarification reduced fouling	–	–	-
	C	–	– best membrane performance was observed on clarified water followed by coagulated water	–	–	-
	D	–	–	–	–	-
	E	–	– CA fouled slightly less than PS	–	–	-

A: impact of water quality   B: impact of pretreatment chemicals   C: impact of pretreatment processes   D: impact of configurations   E: impact of membrane materials

General Conclusions

In addition to the specific impacts on membrane performance summarized above, the following general conclusions from the bench- and pilot-scale results can be made:

- No single measured water quality parameter, such as pH, alkalinity, hardness, total organic carbon (TOC), UV254 absorbance, specific ultraviolet absorbance (SUVA), or silica could be correlated to the extent of membrane fouling. A combination of several factors, or water quality characteristics that are not adequately described by current parameters, may be responsible for membrane fouling.
- Bench-scale studies typically agreed qualitatively with pilot-scale studies, suggesting that the preliminary membrane selection may be best conducted in the laboratory under more controlled conditions.
- SEM and FTIR analysis provided qualitative results of the membrane performance, which further supported the conclusions of flux decline results.
- Size fractionation provided insight into the nature of foulants from various source waters.
- Future study of fractionation should be used as a tool to further investigate the direct impact of fractionation change by varying pretreatment conditions (e.g., using different coagulants, or different coagulant doses) and the corresponding flux decline.
- The effect of pre-chlorination on membrane performance should be evaluated with long-term testing.
- Coagulant chemical and dose should be optimized based on site-specific water quality and pilot-scale testing. Coagulant dose determined based on traditional water quality parameters from conventional jar testing (i.e., turbidity, TOC, etc.) should be verified for optimal membrane performance.

Full-Scale Case Studies

The key findings of the full-scale plant evaluations confirmed pretreatment effects observed in the bench- and pilot-scale studies. The results are summarized below:

- The City of Westminster's Northwest Water Treatment Facility data supported pilot-scale results comparing raw water to clarified water. The specific flux decline rate without pretreatment was more than twice the rate when the membrane system was operated with pretreatment.
- The Ashland Microfiltration Water Treatment Plant data showed that in-line coagulation at a low dose could remove color (seasonal issue) and assist in stabilizing the membrane transmembrane pressure (TMP).
- The Meredith Water Treatment Plant provided data verifying that aluminum chlorohydrate (ACH) performed better than alum for membrane pretreatment. At the same metal mole dose, the specific flux decline rate of a membrane fed with ACH coagulated water was slower than that when alum was used as a coagulant.
- The Waikato Water Treatment Plant data demonstrated the interaction of coagulant and raw water quality. At Waikato, high levels of silica in the raw water combined with aluminum from the coagulant to form kaolinite, [$Al_2Si_2O_5(OH)_4$], which was

determined through membrane autopsy to be the major foulant. Periodic citric acid maintenance cleaning efficiently reduced the membrane inorganic fouling rate.

Cost Analysis

The cost analysis was based on a model originally developed by Pickering and Weisner (1993) and updated with more recent cost information from a membrane survey (Adham et al. 2005). The pilot-scale testing results were used in the cost analysis based on the current or future plant capacity. Each case of the cost analysis is summarized below:

- Both flocculation and sedimentation pretreatment are cost effective for a membrane plant with water quality similar to Chattahoochee River. The pilot-scale testing results at AFCWTP using Chattahoochee River water demonstrated that the normalized specific flux increased by 40% when operated with raw water compared to flocculated water. Based on the cost model, a 96 mgd membrane facility will have a capital cost decrease of 16% when flocculation is used as membrane pretreatment. The capital cost will decrease by 19% when sedimentation is used compared to direct filtration with raw water, based on the pilot results which demonstrated that the normalized specific flux of settled water was 59% higher than raw water.
- Sedimentation is not a cost effective pretreatment for a membrane facility with water quality similar to Seine River. The pilot test results obtained at CIRSEE using Seine River water as water source demonstrated a moderate normalized specific flux increase of 17% with settling compared to raw water. A 14.5 mgd plant will cost 11% more with settled water compared to raw water. The flux increase is not high enough to recover the extra capital cost of a settler. However, with addition of polymer at an optimized dose, the capital cost of the membrane plant with settling and polymer addition will be close to the capital cost of a plant without pretreatment.
- Neither flocculation nor sedimentation is a cost effective pretreatment for a membrane plant with water quality similar to Clackamas River. The pilot testing results at NCCWC using Clackamas River water showed slight normalized specific flux increase, 5%, when the membrane system was operated with flocculated water compared to raw water. The cost model predicted 7% increase in capital cost when a 10 mgd plant selects flocculation without sedimentation as pretreatment for membrane compared to direct filtration. A 10% increase in the normalized specific flux was observed during pilot testing when the system was operated with sedimentation compared to raw water (without pretreatment). The capital cost was predicted to be 15% more with the pretreatment process compared to without pretreatment.

CHAPTER 1
INTRODUCTION

BACKGROUND

Microfiltration (MF) and ultrafiltration (UF) have become accepted water treatment filtration technologies. Since the first MF plant in the United States was built in 1987, over a hundred MF and UF plants have been constructed, with many more in other countries. This growth has been spurred primarily by concern over the ability of conventional treatment to provide adequate protection against disinfection-resistant microorganisms and meet ever-expanding regulatory requirements. In addition, rapid advances in MF and UF equipment and increased competition among vendors have made the processes more cost effective.

Early applications of MF and UF to water treatment relied on cartridge filters to protect the membranes. The success of early installations created an increased interest in MF/UF membranes among utilities with a wide variety of source waters. These included utilities needing disinfection by-product control or removal of natural organic matter (NOM), color, taste and odor, hardness, iron, manganese, arsenic, and other organic and inorganic compounds. These source waters are more challenging for membranes, however. Constituents in the raw water such as high turbidity and/or high organics can lead to membrane fouling. In addition, MF/UF membranes do not, by themselves, provide adequate treatment for the above wide variety of contaminants. In order to treat challenging water supplies, additional processes must be considered for pretreatment to MF and UF, both to maintain the productivity of the membrane process and to meet treatment objectives that are not met by the membranes.

Given the success of the early stand-alone membrane installations, many utilities began to also consider integrating membranes into their existing conventional treatment plants for various reasons, including increasing plant capacity, meeting more stringent regulatory requirements, producing consistent effluent quality and upgrading plant infrastructure. Integrating membranes into an existing treatment plant presents many challenges such as identifying the best location for integration with the treatment train and determining the compatibility of the membrane material and system configuration with the chemicals and infrastructure used in the existing treatment plant.

The successes of newly designed installations on clean source waters are not directly transferable to more challenging water supplies and/or system applications. Insufficient information is available about how the interaction between raw/clarified water constituents and water treatment chemicals will affect membrane flux, backwashing requirements, cleaning frequency, cleaning efficiency, and membrane longevity. Thus, knowledge of the effect of pretreatment processes on MF/UF membranes is critical to the future application of membranes in water treatment.

OBJECTIVE

The objective of this project is to investigate the impact of the interaction among water quality, pretreatment processes, pretreatment chemicals, membrane materials and membrane configurations on the membrane performance. This provided the foundation for a comprehensive understanding that the drinking water community can utilize to best decide when and how to optimize the integration of MF/UF technologies into their existing or new conventional treatment plants.

RESEARCH APPROACH

This project was implemented in four phases: literature survey, conceptual assessment, bench-scale testing, and field evaluation.

Literature Survey

In the first phase, a comprehensive literature review was initiated to update and summarize the current state of knowledge on issues related to the rapidly developing field of MF/UF integration to treatment plants and the fouling mechanism on membrane.

Conceptual Assessment

The conceptual assessment was performed through two workshops. The first workshop focused on setting the overall technical direction for the project, with special emphasis on the bench-, and pilot-scale testing. The second workshop reviewed up-to-date bench-scale and field evaluation data and set the technical direction for completing the project, with special emphasis on the development of the optimization guidance database.

Bench-Scale Testing

The bench-scale testing phase of this project was initiated after the first workshop. The primary focus of bench-scale testing was to gain insight on membrane fouling mechanisms by investigating interaction between water quality, pretreatment chemicals and membrane material. The bench-scale work evaluated several source waters, representing a global geographic distribution. Several membrane materials, representing the common materials currently in use in full-scale facilities, were tested at bench-scale.

Field Evaluation

The field evaluation phase of this project was initiated by the middle of the bench-scale testing phase. This phase included a pilot-scale study and full-scale plant evaluation. The primary focus of the field evaluation was on the interaction between water quality, pretreatment chemicals, membrane materials and configurations to gain insight into operational and design issues that are impacted by membrane pretreatment. Various utilities were consulted to provide experiences regarding the impact of pretreatment processes and chemicals on the membrane performance. This assured the practicality and applicability of the final project outcome. The ultimate goal of the project was to correlate the bench-, pilot-, and full-scale data, and develop a guidance of pretreatment for membrane filtration.

CHAPTER 2
LITERATURE SURVEY

One component of this study was to review the literature regarding the effects of pretreatment on membrane filtration, and prepare an updated summary of this literature. Much has been written about pretreatment for membrane filtration in the past 15 years. Early use of MF and UF in water treatment relied on clean source waters, so the first plants had no pretreatment except cartridge filters to protect the membranes. The success of these installations prompted a close look at membrane filtration by other utilities, including utilities needing disinfection by-product control or removal of NOM, color, taste and odor, hardness, iron, manganese, arsenic, and other organic and inorganic compounds. The first reported membrane facility including coagulation ahead of membrane filtration was in 1998 (Best et al. 2001), but many such plants exist today. Treatment plants are now in operation where membranes are preceded by oxidants (including chlorine, potassium permanganate, and ozone), coagulation and flocculation, sedimentation, lime softening, and activated carbon (Freeman 2001). Literature on pretreatment has been particularly evident in conference proceedings, where the results of pilot studies using a variety of pretreatment strategies for membrane filtration have been reported. Pretreatment has been included for a variety of objectives, such as to improve membrane performance or to provide treatment for parameters, which are not removed by membrane filtration alone (such as arsenic).

Literature on pretreatment for membrane filtration is summarized in 0. The table provides details on the source water used in the study, the type of membrane, the pretreatment used, and the focus of the study. Abbreviations and details about the membranes cited in the table are also identified following the table. In addition, a summary of the major issues identified in this literature is discussed in the following paragraphs. A discussion of the effect of coagulation on membrane performance is provided first, followed by discussions of the effect of activated carbon, and oxidants. An excellent review on the effect of pretreatment on membrane filtration has also recently appeared in the technical literature (Farahbakhsh et al. 2004).

COAGULATION AND FLOCCULATION

The primary use of coagulation and flocculation in water treatment is to destabilize particles to facilitate their removal by sedimentation and granular media filtration. Although sedimentation may reduce the solids loading on membrane filters, destabilization is not necessary for membrane filtration because particles are removed by straining instead of depth filtration. Coagulation is also used for the removal of NOM, color, and inorganic compounds such as arsenic. Removal of NOM is significant because NOM is frequently identified as a primary membrane foulant. Relationships between coagulants, NOM, and membrane surfaces is complex, and both beneficial and adverse effects have been reported when coagulation and flocculation is used as pretreatment to membrane filtration.

A number of studies have demonstrated that coagulation pretreatment can improve membrane performance. Gallagher et al. (2005) found an increase in membrane flux after coagulation with aluminum chlorohydrate and ferric chloride in a pilot study evaluating manganese removal. Braghetta et al. (2001) found that coagulation with ferric salts improved membrane flux by 25 to 50 percent in pilot testing with two natural surface waters. Both pilot studies evaluated ultrafiltration membranes from three manufacturers using an up-flow sludge blanket clarifier for pretreatment. Robert et al. (1999) found that cellulose acetate (CA) ultrafiltration membranes fouled rapidly without pretreatment, but the addition of 21 mg/L of ferric chloride as pretreatment essentially eliminated fouling. Jack and Clark (1998) conducted a pilot study using lake water and CA ultrafiltration membranes. Pilot operations had been effective for several separate experiments until one experiment began experiencing rapid fouling. An addition of 14-mg/L ferric chloride restored the pilot unit to near normal operation. Laîné et al. (1989) found a significant improvement in membrane flux in batch tests of reconstituted lake water after coagulation. Terminal flux through a hydrophilic regenerated cellulose membrane increased from 20 percent of initial flux without coagulation to 60 percent with coagulation with 60 mg/L of polyaluminum chloride. Humbert et al. (2005) evaluated the use of coagulants and ion exchange resins, including MIEX, and found pretreatment with ion exchange resins did not improve membrane flux, although an improvement was observed with coagulants.

Coagulation has had an adverse impact on membrane performance in other studies. Kader (2005) found that coagulation with ferric chloride for arsenic removal caused an increase in the rate of transmembrane pressure increase, although the effect was relatively minor. Liu et al. (2005) found that ferric chloride could combine with natural organic matter from the source water to form complexes that could foul the membrane, and that this effect was related to the low alkalinity of the feed water. A switch to a polyaluminum chloride (PACl) product that did not consume a significant amount of alkalinity appeared to perform better than ferric chloride. During pilot testing of Ohio River water with polypropylene microfiltration membranes with 1 to 5 mg/L of alum or PACl, a sharp decline in specific flux was observed with in-line coagulation compared to a gradual decline in specific flux without coagulation (Shrive et al. 1999). Addition of 3 mg/L of ferric chloride during pilot testing of polypropylene microfiltration membranes on surface water in California resulted in accelerating the buildup of transmembrane pressure (Karimi et al. 1999). Shorney et al. (2001) conducted a pilot study of membrane filtration following coagulation with ferric sulfate. Four membrane systems were evaluated with coagulant doses ranging from 0 to 15 mg/L. In-line coagulation was practiced with 4 to 16 minutes of coagulant contact time prior to the membranes. Mechanical and operational problems made it difficult to assess the impact of coagulation on performance of two of the systems, but fouling was worse with coagulation in the other two systems. Schäfer et al. (2001) compared MF and UF processes with and without ferric chloride pretreatment for filtration of a solution containing Suwannee River humic and fulvic acids at a concentration of 5 mg/L as DOC. Coagulation with ferric chloride significantly increased the fouling of the MF and UF membranes.

Differences in performance after coagulation have also been observed with polymers as coagulants. Burns et al. (2005) observed worse performance after the direct injection of cationic polymer in the membrane feed water. However, improvements after feed water pretreatment with polymers has been observed in other studies, as will be seen later in this report.

Some of the conflicting results may be due to differences in the operating conditions selected. Howe and Clark (2002a) examined the effect of coagulant dose in detail using five surface waters and three membrane materials, using coagulant doses ranging from an "ineffective" dose to the dose necessary for enhanced coagulation. The lowest alum doses caused an increase in fouling in a significant number of cases, and the highest doses always improved membrane performance. Similar results were obtained by Judd and Hillis (2001) and Fu and Dempsey (1997). Judd and Hillis (2001) examined microfiltration of coagulated natural waters with a range of doses of ferric salts. Fouling increased slightly at a low ferric dose but overall membrane performance improved at higher ferric doses. Fu and Dempsey (1997) filtered coagulated 10 mg/L solutions of Dismal Swamp fulvic acids through a polysulfone UF bmembrane in a stirred cell with alum doses ranging from 30 to 210 mg/L. Low alum doses caused an increase in the rate of fouling, but the highest doses improved performance compared to the raw water. Using PACl and CA membranes operated in cross-flow mode, however, Bian et al. (1999) found the opposite effect. Optimal flux improvement was seen at a low PACl dose, and the highest PACl doses appeared to have as much flux decline as uncoagulated water. The authors found that optimal conditions for conventional treatment and humic substance removal was at higher aluminum doses, but that the optimal coagulant dose for membrane operation was lower than that for conventional treatment or humic substance removal. The authors hypothesized that higher coagulant doses produced sticky aluminum micro-flocs, which formed a cake layer on the membrane.

POWDERED ACTIVATED CARBON

Powdered activated carbon (PAC) is frequently used in conventional water treatment plants for the control of taste and odor or removal of synthetic organic chemicals, and is capable of removing NOM to control disinfection by-product formation. Several treatment plants incorporating PAC into a recirculation loop in cross-flow UF membranes are in operation. These systems utilize PAC for the removal of dissolved natural or synthetic compounds, and then use the membranes to separate the PAC, as well as microorganisms and other particulate matter, from the finished water. The combination of PAC and UF in a single process can be successful in removing organics (Adham 1993). Sophistical models have been developed to predict the performance of combined PAC/UF systems (Adham 1993, Campos et al. 2000a, Campos et al. 2000b).

NOM has been implicated in researches to be an important foulant of MF/UF membranes. PAC can be highly effective for removing NOM. Thus, it might be expected that pretreatment with PAC could yield substantial reductions in fouling. In cross-flow systems, PAC addition has been observed to reduce the rate of membrane flux decline, but have little effect on the extent of irreversible fouling (Adham 1993). The PAC could affect the rate of flux decline through a variety of mechanisms, including a scouring effect on the cake that forms on the membrane surface or an increase in porosity of the surface cake, similar to the effect of body feed in diatomaceous earth filtration. The ability of PAC to adsorb NOM, however, appears to have little effect on membrane performance. Laîné et al. (1990) applied a dose of 250 mg/L of PAC to Lake Decatur (Illinois) water, which removed 85 percent of the total organic carbon (TOC) but had little impact on flux in dead-end filtration. Similarly, Carroll et al. (2000) applied a 100 mg/L dose of PAC to Morrabool River (Australia) water, which removed 68 percent of the dissolved organic matter (DOC) but had no impact on flux in dead-end filtration. Lin et al. (2001) found that PAC made membrane fouling worse. Thus, the causes of membrane fouling in natural waters are more complex that just the presence of NOM, and processes that remove NOM cannot be guaranteed to have a positive effect on membrane performance. The effect of PAC on membrane performance must be evaluated in conjunction with other treatment chemicals to develop a unified understanding of membrane fouling.

Zhang et al. (2003) found that adsorbents other than carbon might have a beneficial effect on membrane performance. They found that heated iron oxide particles (HIOPS) could form a porous layer on the membrane that adsorbed foulants from the feed water and improved membrane flux.

CHEMICAL OXIDANTS

Little information is available in the literature on the effect of oxidants on membrane flux, backwashing frequency, or cleaning frequency. Schneider et al. (2001) reported on studies involving three source waters and several oxidants for the removal of manganese. During oxidation, manganese is oxidized from a soluble form to an insoluble form, which can then be filtered by membrane filtration. The studies noted no adverse affects on the membranes, and did not report any changes in flux or cleaning frequency.

An important issue regarding the use of oxidants before MF/UF membranes is the chemical compatibility, which affects material longevity rather than performance parameters such as flux, backwashing frequency, or cleaning frequency. Some membrane materials are resistant to common oxidants while others are not. Freeman (2001) noted that plants employing oxidants for pretreatment use membranes fabricated of polysulfone (PS), polyvinylidene fluoride (PVdF), or a resistant proprietary material, which have high chemical resistance. Other common materials, such as polypropylene (PP) and CA, are more easily damaged by oxidants, although resistance depends on the oxidant as well as the membrane material.

CHAPTER 3
BENCH-SCALE

The objective of the bench scale experiments, conducted in the University of New Mexico (UNM) Environmental Engineering Laboratory, is to evaluate the impact of raw and pretreated water on MF and UF membrane performance.

The bench-scale studies used five natural river waters collected by participating utilities. Some utilities also sent water that was clarified at the utility for comparison of results with water clarified at the laboratory. Simultaneously, pilot studies were performed at the utility site with same source water and similar membranes. Water was coagulated and clarified in the laboratory in accordance with the coagulant dose used at the corresponding pilot plant or full-scale facility. For some experiments, membrane feed water was fractionated for further investigating the characteristics of fouling material. The membranes used for the experiments were analyzed using scanning electron microscopy (SEM) and Fourier transform infrared (FTIR).

MATERIALS AND METHODS

Water Preparation

The water was shipped to UNM in 5-gallon collapsible low-density polyethylene (LPDE) containers stored in plastic picnic coolers, which were refrigerated upon arrival. The coolers were filled with packaged ice to keep the water at the same temperature and avoid deterioration in water quality. Preservative was not added to the raw water for storage. The amount of water shipped was based on the list of experiments to be accomplished with the source water. Prior to membrane flux experiments, the water was taken out of the refrigerator the evening prior to use and stored in the dark to allow the water to reach room temperature while avoiding any microbial growth. The raw water container was rolled over a platform five times to homogenize the water prior to use. Raw water was analyzed for TOC and UV_{254} absorbance upon arrival and periodically during testing to monitor water quality. Table 3.1 gives a brief description of the raw water tested for the project. It took an average of two months to do the experiments and analysis for each of the source waters.

Three types of feed water were tested in the bench scale studies: (1) raw water, (2) water clarified in the laboratory, and (3) water clarified at the water utility.

Table 3.1
Source waters used in bench-scale testing

Source Water	Location	WTP	Pretreatment
Medina River	Texas, USA	Bexar Met WTP	Ferric sulfate
Chattahoochee River	Georgia, USA	Atlanta Fulton County WTP	Alum
Seine River	France	Vigneux-sur-Siene WTP	Ferric chloride, alum, PAC
Clackamas River	Oregon, USA	Sunrise WTP	Ferric sulfate, alum, ACH, chlorine
Waikato River	New Zealand	Waikato WTP	Alum

Coagulated Water Preparation

The raw water was removed from the refrigerator and allowed to reach room temperature before it was coagulated. All equipment were cleaned and rinsed with deionized (DI) water prior to use. The raw water was shaken well before pouring in the jars. Coagulation was conducted using a Phipps and Bird 6-position laboratory jar tester with 2-L square jars. Six jars were used for coagulation at a time, with 2 L of water in each jar. Stock solution of the coagulants used was prepared ahead of time. The targeted dose was pipetted directly into the mixing jars or measured in a plastic cup to facilitate rapid addition. The water was mixed at the highest speed while adding the coagulant dose for flash mix. The mixer speed was reduced to 60 rev/s for flocculation, which provided a velocity gradient (G value) of 60 s^{-1}. The flocculation time was 30 minutes. The mixer was turned off and water was allowed to settle for 30 min. Only 1.5 L of water was taken out of each jar, leaving the settled flocculated particles in the jar. The clarified water was stored in cubitainers and refrigerated.

The dose of the coagulant was the same as in the corresponding pilot plant. The various pretreatment chemicals used for each source water are shown in Table 3.1. For most source waters, coagulant was provided by the utility to provide consistency between bench and pilot scale testing. Alum (as $Al_2(SO_4)_3{\cdot}18H_2O$) and ferric chloride (as $FeCl_3{\cdot}6H_20$) were used as coagulants in most of the source waters. For the Medina River, coagulation at the utility was with the commercial grade chemical but coagulation in the laboratory was with reagent-grade ferric sulfate ($Fe_2(SO_4)_3{\cdot}9H_20$). Coagulant doses were based on the molar quantity of the metal ion (Al^{3+} or Fe^{3+}) so that the molar dose of the chemical was always consistent with the pilot plant studies. Some waters were tested with more than one coagulant to evaluate the effect of different coagulants with the same source water. In studies where more than one coagulant was used in the bench scale studies (Seine and Clackamas Rivers), the dose of various chemicals was consistent on molar basis; the Fe molar dose was calculated on the basis of the dose given by the utility and the corresponding Al molar dose was calculated and used for the alum coagulation. For the Seine River, the solutions of alum and ferric chloride shipped from the utility had different densities and percent concentrations. The dose calculated was 70 mg/L of alum solution and 40 mg/L of ferric chloride solution to provide the same molar concentration.

The Seine River experiments also included some experiments with PAC, using the dose (2 mg/L) and PAC provided by the utility. PAC is used in Seine River water for taste and odor removal. In previous research, PAC has improved flux in some source waters and increased fouling in others. Table 3.2 describes the chemicals and doses used for each source water. Clackamas River was coagulated with aluminum chlorohydrate (ACH) in addition to alum and ferric chloride. Please note that ACH is one form of the PACl.

Table 3.2
Chemicals and doses used with each source water

Source water	Chemical	Dose	Chemical	Dose
Medina River	Ferric sulfate	6.87 mg/L as Fe		
Chattahoochee River			Alum	0.68 mg/L as Al
Seine River	Ferric chloride*	7.09 mg/L as Fe	Alum*	3.19 mg/L as Al
Clackamas River	Ferric sulfate*	0.94 mg/L as Fe	Alum*, ACH	0.45 mg/L as Al
Waikato River			Alum	2.27 mg/L Al

* These chemicals were shipped as solutions from the utility

Fractionated Water Preparation

Some bench scale tests included fractionation of the feed water by prefiltration through various size membranes. The purpose of the size fractionation was to understand the fouling mechanism for the source waters in relation to the size of various components in the feed waters. Each water sample was fractionated to exclude specific molecular weight cutoff (MWCO) components listed below:

- Not fractionated
- 1 μm fractionated
- 100 kilodalton (kDa) fractionated
- 10 kDa fractionated

Fractionation was carried out immediately prior to membrane flux experiments to provide quality control and to prevent the water from being contaminated. Glass fiber filters (Gelman Sciences, type A/E) were used for the 1 μm fractionation. The filtration volume was limited to 400 mL of water per filter to avoid overloading the membrane and to prevent removal of particles smaller than the nominal pore size.

The 10 kDa and 100 kDa fractionation was performed using regenerated cellulose ultrafiltration membranes (Amicon YM10 and YM100, respectively). The membranes were 63.5 mm in diameter and used with a pressurized filter module. The membranes were soaked in DI water for 30 minutes to remove any organic preservatives from the surface of the membrane. In addition, DI water was filtered through the membrane for 30 minutes to normalize the flux and to further remove any preservatives from the membrane. The feed water was passed through the membrane at 70 kPa (10 psi) for the 100-kDa membranes and 413 kPa (60 psi) for the 10-kDa membranes. Additional details on the membrane filtration procedures are provided in the following section.

Membrane Materials

Bench-scale tests were performed with PS ultrafiltration and PVdF microfiltration membranes. These membranes were selected to match the type of membrane being used at the corresponding pilot or full-scale plant and to represent the commercially available MF/UF membranes for drinking water applications. A MF/UF membrane knowledge base survey (Adham et al. 2005) showed the trend of the membrane materials used in the water industry. Table 3.3 shows the membranes materials and configurations that are commonly employed at water treatment plants. The materials can be categorized to four groups: cellulose acetate (CA), polysulfone (PS) or polyethersulfone (PES), polyvinylidene fluoride (PVdF), and polypropylene (PP). Among these groups, CA and PP membranes are the earliest products, while PVdF and PS/PES have gained popularity as base material for the majority of available membrane products. This project chose PS ultrafiltration and PVdF microfiltration to represent the membrane materials employed in drinking water field.

Bench scale experiments were carried out using both hollow fiber and flat sheet membranes to get a understanding about fouling of hollow fiber membranes and to evaluate the differences between hollow fiber and flat sheet membranes. Table 3.4 describes the membranes used with each source water. Characteristics of the membranes are shown in Table 3.5 for the PVdF membranes and Table 3.6 for the PS membranes.

Table 3.3
Membrane materials commonly employed at water treatment plants

Membrane Vendor	Material	Configuration
Aquasource	Cellulose Acetate (UF)	Pressurized
	Polysulfone (UF)	Pressurized
Hydranautics	Polyethersulfone (UF)	Pressurized
Koch	Polysulfone (UF)	Pressurized
Norit	Polyethersulfone (UF)	Pressurized
Pall	Polyvinylidene fluoride (MF)	Pressurized
USFilter	Polypropylene (MF)	Pressurized, submerged
	Polyvinylidene fluoride (MF)	Pressurized, submerged
Zenon	Polyvinylidene fluoride (UF)	Submerged

Table 3.4
Membranes used with each source water in the bench scale tests

Source water	Type	Material	Flat sheet Source	Hollow fiber Source
Medina River	UF	Polysulfone	Aquasource	--
Chattahoochee River	MF	PVdF	Millipore	US Filter
Seine River	UF	Polysulfone	Aquasource	Aquasource
Clackamas River	UF	Polysulfone	Aquasource	Aquasource
Clackamas River	MF	PVdF	Millipore	US Filter
Waikato River	UF	Polysulfone	Aquasource	Aquasource
Waikato River	MF	PVdF	Millipore	--

Table 3.5
Bench-scale PVdF membrane specifications

Parameter	**Flat Sheet**	**Hollow Fiber**	**Hollow Fiber**
Manufacturer	Millipore	US Filter	US Filter
Catalog Number	Durapore GVWP 04700	PVdF1	PVdF2
Membrane Material	PVdF	PVdF	PVdF
Flow Direction	Downward	Outside-In	Outside-In
Maximum TMP	3.45 bar (50 psi)	1.5 bar (22 psi)	1.48 bar (21.7 psi)
Fiber Inside Diameter (mm)	--	0.31	0.5
Fiber Outside Diameter (mm)	--	0.65	0.8
Membrane thickness (μm)	125	170	150
Membrane surface area (cm^2)	17.3	13.0*	12.9*
Nominal Pore Size (μm)	0.22	0.1	0.1
Specific clean water flux ($L/m^2 \cdot h \cdot bar$)	3000	200	147
Surface Characteristics	Hydrophilic	Hydrophobic	Hydrophilic
Contact angle	40°-50°	--	--

*when hollow fiber module had a single fiber

Table 3.6
Bench-scale PS membrane specifications

Parameter	**Flat Sheet**	**Hollow Fiber**
Manufacturer	Aquasource	Aquasource
Membrane Material	polysulfone	polysulfone
Flow Direction	Downward	Inside-out
Maximum TMP	1.5 bar (22 psi)	1.5 bar (22 psi)
Fiber Inside Diameter (mm)	--	0.93
Fiber Outside Diameter (mm)	--	1.67
Membrane thickness (μm)	--	370
Membrane surface area (cm^2)	17.3	18.1*
Molecular Weight Cut-off (Da)	100,000	100,000
Specific clean water flux ($L/m^2 \cdot h \cdot bar$)	600	400
Surface Characteristics	Hydrophilic	Hydrophilic
Contact angle	35°-50°	--

* when hollow fiber module had a single fiber

An effort was made to investigate the relative hydrophobicity of each of the membrane products used in this research. Unfortunately, a quantitative comparison of hydrophobicity could not be made. One common measure of hydrophobicity is contract angle, in which a droplet of water is placed against a membrane surface and the tangent angle between the air, water, and membrane surface is measured through the liquid phase. The tendency of the water to spread against hydrophilic surfaces results in a low contact angle, whereas hydrophobic surfaces tend to have a high contact angle. Figure 3.1 shows the typical contact angle measurement apparatus and the contact angles for hydrophilic and hydrophobic surface.

Contact angle measurements can be obtained for flat surfaces, but are unavailable for configurations such as hollow fiber membranes. Contact angle measurements for PS membranes typically range from 60 to 70 degrees with variation due to surface chemistry modification. The manufacturer of the flat sheet PVdF membrane reported that contact angles of 40 to 50 degrees are typical for that product, depending on which pore size is tested, because the contact angle can be affected by the roughness of the material.

US Filter provided more than one type of PVdF hollow fiber membrane for this project. Although each membrane was principally made of PVdF, each was made by a proprietary process that had the potential to affect the hydrophobicity of the material. As a result, US Filter referred to one product as a "hydrophobic PVdF" (PVdF1) and another as a "hydrophilic PVdF" (PVdF2). It should be noted that Millipore, the supplier of the flat sheet membranes for this study, also market a hydrophilic PVdF and a hydrophobic PVdF flat sheet membrane, where the hydrophilicity has been altered by modification of the surface chemistry. US Filter characterizes the membranes by the ability to adsorb water and by the permeation that can be generated by a membrane that has not been "wetted out." As a result, the characterizations as hydrophilic or hydrophobic are based on relative comparisons between products which cannot be used to provide a quantitative basis for the relative hydrophobicity of the different materials used in this study.

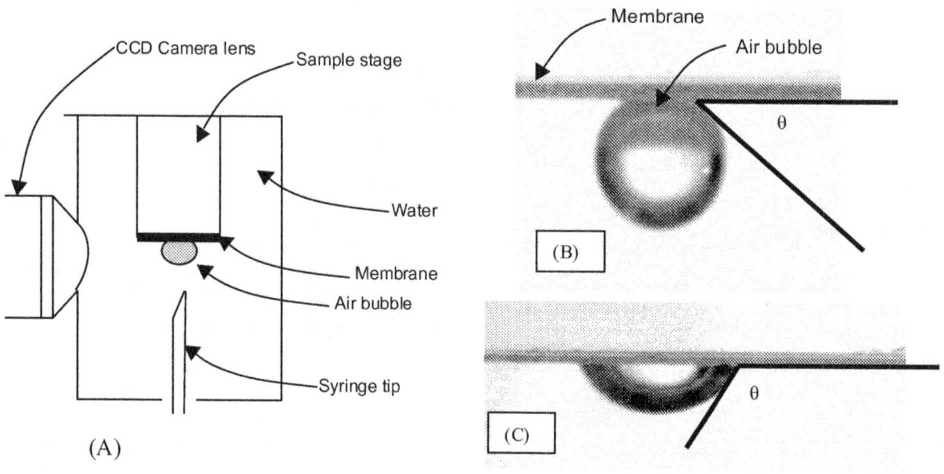

Figure 3.1 Captive bubble contact angle measurements for determination of hydrophobicity: (A) Contact angle measurement apparatus, (B) Hydrophilic surface (low contact angle), and (C) Hydrophobic surface (high contact angle)

Membrane Filtration Procedures

Flat Sheet Membrane Setup

The PS ultrafiltration membranes were provided by Suez Environment (Aquasource) and were shipped at wetted condition. These membranes were kept wet in DI water at all times and were stored in the refrigerator to avoid deterioration. The membranes were soaked in 1 percent bisulfite to prevent degradation. The PS membranes were provided in sheets and cut to the appropriate size.

The PVdF microfiltration membranes were purchased from Millipore (GVWP 04700) and no initial preparation was necessary. The membranes were wetted with DI water first and care was taken for both PS and PVdF membranes so that the shiny side of these membranes faced the flow of water. Flat sheet experiments were performed using 47 mm diameter membrane discs. These discs were used with a pressurized module at 1 bar (100 kPa or 14.5 psi) and DI water was passed through the membrane for 30 minutes for normalizing the flux. A schematic of the bench scale setup is shown in Figure 3.2, and a photograph is shown in Figure 3.3.

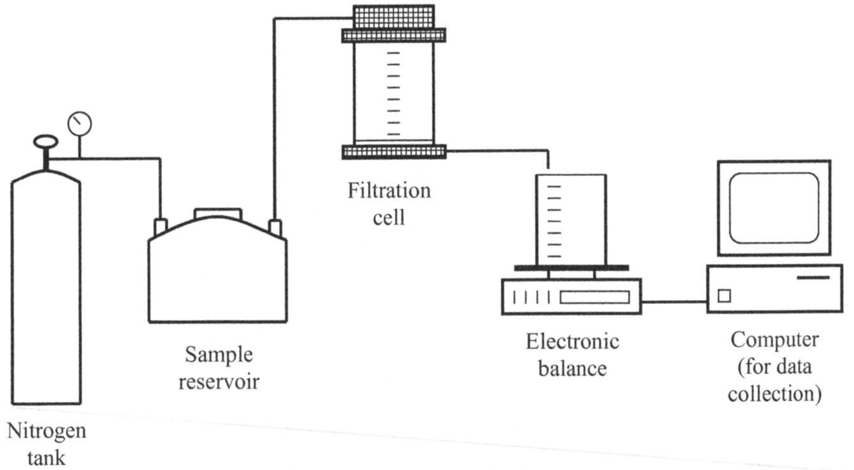

Figure 3.2 Schematic of membrane filtration equipment setup

Figure 3.3 Bench-scale flat sheet experimental setup

Hollow Fiber Membrane Module Fabrication

Hollow fiber membrane modules were made at UNM with tubing and epoxy. The first modules were made with 9 membrane fibers, but the permeate flow through the module was too high, so subsequent modules only used 1 membrane fiber per module. The hollow fiber membrane modules were used at a pressure of 70 kPa (10 psi) because higher pressures tended to cause the module to leak. A similar approach was adopted for hollow fiber experiments as the flat sheet membrane experiments, in which DI water was passed through the membrane module for 30 minutes for flux normalization. The bench-scale experimental setup for hollow fiber membranes is shown in Figure 3.4.

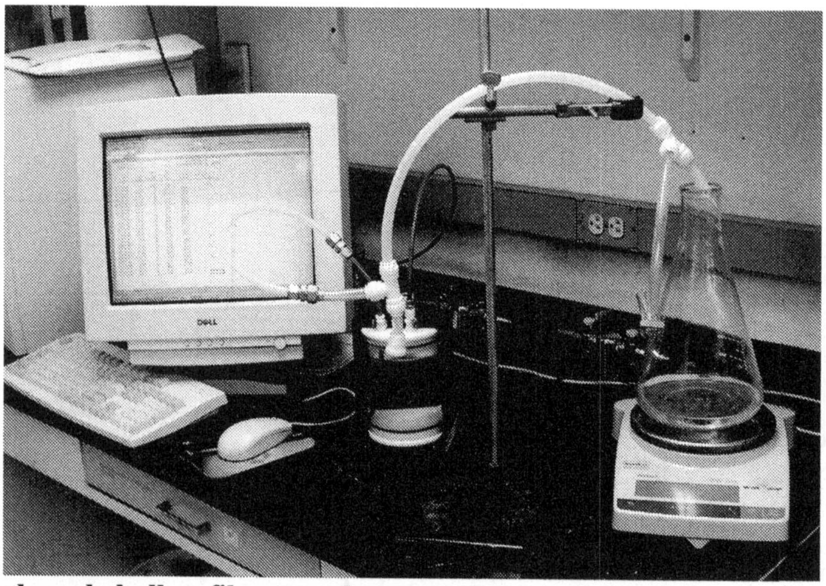

Figure 3.4 Bench-scale hollow fiber membrane setup

Hollow fiber modules were fabricated using procedures developed for the project; care was taken in handling the membrane fibers and protective gloves were used in potting the membrane in the tube. The tubes used were translucent soft PVC tubes. After putting the membrane fiber through the tube, a T-joint was attached to both the ends and the membrane fiber was allowed to extend outside of the T-joint. A short piece of tubing (about 2.5 cm) was attached on both the ends of the T-joints. This tubing was plugged with silica epoxy and second coat of fast setting epoxy. One end of the membrane fiber was allowed to extend out of the epoxy potting and the other end was plugged by epoxy setting. The feed water inlet was located at the T-joint of the end where the membranes were blocked. The water was collected in a container on a balance, and flux was calculated by a computer. Before starting the experiment, the module was pressurized for checking leaks, and a new module would be built when leaks were detected. All hollow fiber tests were performed in outside-in mode.

Long Term Hollow Fiber Test Setup

A major difference between bench scale and pilot plant or full-scale operation is that in bench scale tests it is difficult to incorporate backwashing, which is an important component in membrane filtration. The other significant difference is that bench scale tests were done at constant pressure condition and the pilot plant tests are generally done at constant flux conditions. A long-term module was developed to examine the effect of backwashing at bench scale. These modules were operated over a 24-hour period and included a filtration cycle that had a 58-minute filtration phase followed by a 2-minute backwash phase at the same pressure (70 kPa), and which was followed by a 10-second rinse. The filtration and backwash cycle was controlled by a timer that opened and closed pinch valves in the feed, backwash, and permeate lines. A photograph of the long-term bench-scale membrane filtration setup is shown in Figure 3.5.

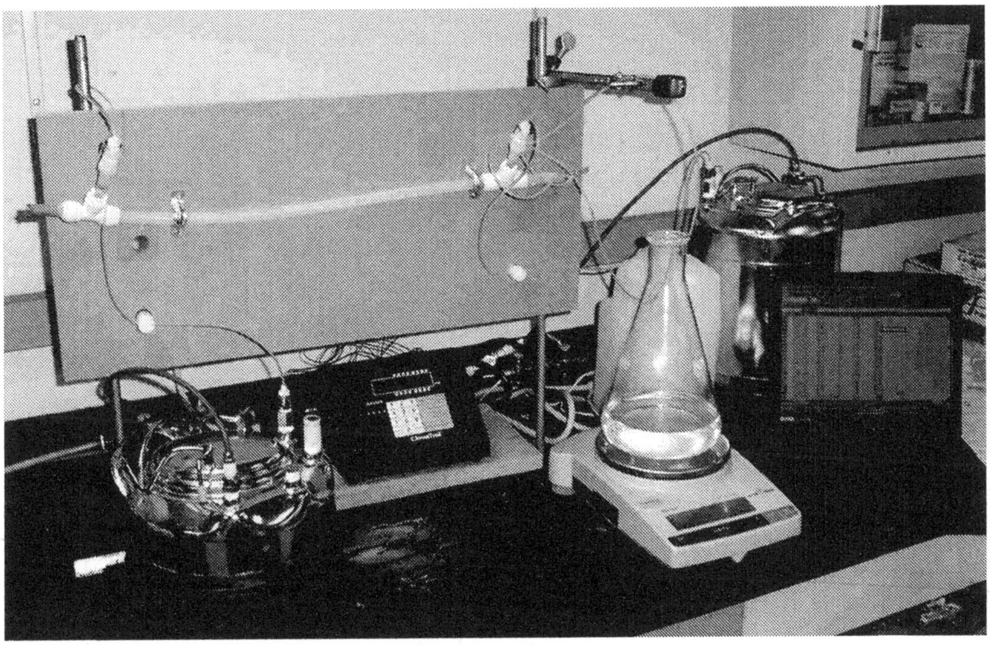

Figure 3.5 Long-term hollow fiber membrane setup

Membrane Flux Tests

Membrane flux tests were performed on raw, clarified, and fractionated waters. The membrane filtration procedure adopted in this research was based on previous research (Howe and Clark 2002b).

Membrane Filtration Procedure

As described earlier, the flat sheet membrane experiments were performed with the unstirred batch filtration cell using 47 mm membrane discs (Amicon Model 8050, Millipore Corporation, Bedford, MA). The fractionation procedure (100 kDa and 10 kDa), however, used a filtration cell that held a 63.5 mm diameter membrane discs (Amicon Model 8200, Millipore Corporation, Bedford, MA). These filtration cells were connected to the pressurized reservoir containing sample water. A 4-L stainless steel reservoir was used for DI water and a smaller plastic reservoir was used for the feed water sample. Each sample run was preceded by filtering DI water for at least 30 minutes through the PS and PVdF membranes. The same amount of water, approximately 800 mL, was passed through the membrane for each experiment. A nitrogen tank was used for pressurizing the reservoir. The pressure was monitored with a 0-400 kPa digital gage connected on the nitrogen tank. The filtration cell was placed over an electronic balance so that the permeate could be collected in a clean 2-L beaker placed on the balance; flux was determined by weighing the permeate at timed intervals. The balance (Mettle Toledo PB3002-S) was connected to a computer that recorded mass from the balance at the timed intervals, calculated flux, and plotted a graph of flux decline in an Excel spreadsheet. The sample water flux was corrected for temperature and pressure, and normalized against clean water flux using the equations described in the following section. The temperature at which these experiments were performed ranged from 20 to 24°C.

Membrane Flux Normalization

The definitions of calculated parameters commonly used in evaluating the performance of membrane are presented in this section.

Filtrate Flux. The permeate flux is the product water flow divided by the surface area of the membrane. Permeate flux is calculated according to the following equation:

$$J = \frac{Q}{S} \tag{3.1}$$

where J = filtrate flux ($L/h{\cdot}m^2$)
Q = filtrate flow (L/h)
S = membrane surface area based on outside diameter (m^2)

Instantaneous Specific Flux. The term specific flux is used to refer to filtrate flux that has been normalized to the transmembrane pressure (TMP).

$$J_S = \frac{J}{\Delta P} \tag{3.2}$$

where J_s = instantaneous specific flux ($L/h{\cdot}m^2{\cdot}bar$)
ΔP = transmembrane pressure (bar)

Temperature Adjusted Specific Flux. The specific flux was standardized for temperature variation according to the equation below. The specific flux in this equation can be substituted by flux for temperature adjusted flux calculation (MWH 2005).

$$J_{sm} = J_S \times 1.03^{(T_S - T_M)} \tag{3.3}$$

where J_{sm} = temperature corrected specific flux ($L/m^2{\cdot}h{\cdot}bar$)
T_S = standard temperature (20°C)
T_M = measured temperature (°C)

Normalized Specific Flux. The specific flux was normalized against the clean membrane flux with DI water. The clean membrane flux obtained initially was also standardized using Equation 3.3. In order to make the flux comparisons among pilot-scale tests with different operating conditions consistent, the membrane flux was compared as normalized net specific flux, which also counted in the feedwater recovery and productivity factor. Although backwash was not employed in the bench-scale tests, the normalized specific flux is referred to as normalized net specific flux in the figures.

$$J_{std} = \frac{J_{sm}}{J_{sm0}} \tag{3.4}$$

where J_{std} = normalized specific flux (dimensionless)
J_{sm0} = clean membrane specific flux with DI water (L/m^2·h·bar)

Specific Volume. The normalized net specific flux was plotted as a function of the specific volume. The specific volume was calculated by dividing the cumulative volume of water filtered by the area of the membrane surface.

$$V_{sp} = \frac{V}{S} \tag{3.5}$$

where V_{sp} = specific volume (L/m^2)
V = cumulative volume (L)

Analytical Methods

The instruments and methods used for analysis along are shown in Table 3.7.

Table 3.7
Analytical methods and instruments for bench scale testing

Analysis	Instrument/method	Standard Method
pH	Orion Model 720A	4500H+ (Electrometric)
Total silica	Varian (Cary 50 Conc) UV-Visible spectrophotometer	4500-Si D
UV_{254}	Varian (Cary 50 Conc) UV-Visible spectrophotometer	5910
TOC	Tekmar Dohrmann (Phoenix 8000) UV-Persulfate TOC Analyzer	3110 C
Alkalinity	Titration with HCl	2320
Hardness	Titration with EDTA	2340 C
Turbidity	Hach Ratio Turbidimeter	2130
Temperature		2550
FTIR	Bruker Tensor 27 FTIR	
SEM/EDS	Scanning Electron Microscope (JEOL 5800LV SEM) attached with an Oxford Analytical ultra-thin window EDS and an Oxford Isis 300 X-ray analyzer	

Scanning Electron Microscopy

The SEM laboratory contains a JEOL 5800LV SEM, equipped with secondary and backscattered electron and cathodoluminescence (CL) imaging detectors. The instrument is a "low pressure" microscope that can operate at both high and low vacuum. The low vacuum mode allows imaging and analysis of samples that do not have to be dried and coated beforehand. This is particularly useful for handling "wet" samples such as soil, sandstone, and clay. An Oxford Analytical ultrathin-window energy dispersive spectroscopy (EDS) and an Oxford Isis 300 X-ray analyzer are attached to the microscope. The stage can accommodate specimens up to 6" across and has 5 motorized axes with full control from the Oxford computer for automated analyses. The analytical software provides semi- to fully quantitative EDS analysis with high-resolution image acquisition, processing, and analysis. Advanced software capabilities of the Isis include phase distribution analysis based on chemical data from the EDS coupled with image data. The CL detector uses a parabolic-mirror light collecting element connected to a high-sensitivity photomultiplier tube, which provides good spectral response across the visible range into the near UV and infrared ranges. SEM analysis performed on the fouled membrane was done after coating the membranes with gold-palladium. A Danvon 510 sputter coater was used to coat the samples.

Attenuated Total Reflectance Fourier Transform Infrared Spectrometry

Attenuated total reflectance Fourier transform infrared (ATR-FTIR) spectrometry was used to examine the chemical composition of clean and fouled membrane surfaces. The instrument used (Bruker Tensor 27 ATR-FTIR) is a single signal bounce instrument and uses a diamond waveguide. Membranes were placed face down on the waveguide and held in place with calibrated press to get the infrared spectra of each membrane.

Silica

A UV-Visible spectrophotometer (Varian Cary 50 Conc) was used for silica, which was the same instrument used for UV_{254} analysis. The standard method used for silica analysis was Standard Methods 4500-Si D. (Molybdosilicate Method). Sample raw waters and fractionated water samples were stored in plastic bottles (cleaned and acid washed) to avoid any silica leaching from glass vials.

Total Organic Carbon

Standard Methods 3110 C (Persulfate-Ultraviolet Oxidation Method) was used for TOC analysis using a UV-persulfate total organic carbon analyzer (Tekmar Dohrmann Phoenix 8000). 40 mL amber vials were used to store the sample from different fractionations of various source waters. The samples vials were refrigerated prior to analysis avoid any microbial growth.

Any glassware used for TOC analyses, including the beakers used to collect membrane permeate, were cleaned with soap and water, acid washed, and heated to a temperature of 550°C in a muffle furnace to eliminate organics from the glassware.

RESULTS

This section presents the results of the bench-scale tests. As noted earlier, five source waters were tested at the bench-scale. For each test, a predetermined amount of water was filtered; the loss of net specific flux at the end of the filter run (i.e., the flux decline) was used as a measure of membrane performance. The section is organized around five main sections.

First, the water quality of each of the source waters is presented. Second, the effect of source water on membrane performance is presented, with separate subsections discussing the performance of PS and PVdF membranes. Third, the effect of pretreatment on membrane performance is presented, again with separate subsections discussing the performance of PS and PVdF membranes. Fourth, the effect of fractionation on membrane performance is presented. Finally, a comparison of bench scale results using flat sheet membrane and hollow fiber membranes is presented.

Raw Water Quality

Table 3.8 gives the water quality data of the source water used for this AwwaRF study. The Medina and Seine Rivers had the highest hardness and alkalinity of the source waters, and the Chattahoochee, Clackamas, and Waikato Rivers had relatively low hardness and alkalinity. The pH values of the source water ranged from 6.9 to 8.4. TOC ranged from 0.89 mg/L to 3.01 mg/L (in Clackamas River and Seine River, respectively), and UV_{254} absorbance values ranged from 0.031 to 0.069 (in Chattahoochee River and Waikato River, respectively). The specific ultraviolet absorbance (SUVA, UV absorbance divided by TOC) value ranged from 2.3 in the Seine River, suggesting a lower proportion of aromatic or hydrophobic composition of the NOM, to 7.5 in the Clackamas River, suggesting a relatively higher proportion of aromatic or hydrophobic composition of NOM. The Waikato and Clackamas Rivers had higher concentrations of silica and lower alkalinity than the other source waters.

Table 3.8
Raw water quality data

Parameter	Medina River	Chattahoochee River	Seine River	Clackamas River	Waikato River
pH	8.4	6.9	7.2	7.1	7.5
Alkalinity (mg/L as $CaCO_3$)	219	25	215	10	40
Hardness (mg/L as $CaCO_3$)	248	18	236	23	55
Turbidity (NTU)	4.6	6	10-15.2	5.6	12
TOC (mg/L)	1.36	1.23	3.01	0.89	1.91
UV_{254} (cm^{-1})	0.054	0.031	0.068	0.067	0.069
SUVA (L/mg·m)	3.9	2.5	2.3	7.5	3.6
Silica (mg/L as SiO_2)	15.4	6.40	2.81	19.6	33.7

Effect of Source Water on Membrane Flux

This section provides a comparison of fouling between the source waters. The bench scale tests used consistent operating conditions and therefore a comparison of the flux decline curves can be accomplished. In addition, water quality data such as TOC, silica, and UV_{254} absorbance were examined to determine if they provided additional insight into the fouling mechanisms.

The fouling of four raw source waters using the PS ultrafiltration membranes is shown in Figure 3.6. The raw Medina River water had the least fouling, and the Seine River had the most fouling. The Medina River had the highest pH, alkalinity, and hardness, but intermediate values for TOC, UV_{254} absorbance, SUVA, and silica concentration. The Seine River has the highest TOC, a low SUVA value, and low silica concentration, whereas the Clackamas River has low TOC, high SUVA value, and high silica, and the Waikato River had moderated TOC and SUVA, and high silica. The fact that these three waters had considerably higher fouling than the Medina River suggests that not a particular measured water quality parameter (TOC, SUVA, or silica) was the predominant cause of fouling, but the combination of several factors may be important. Although the Medina River had the highest alkalinity and hardness, it is unlikely that these factors contributed to the low fouling nature of this water.

The fouling of three source waters using PVdF microfiltration membranes is shown in Figure 3.7. The raw Clackamas River water had low fouling and the Chattahoochee and Waikato Rivers had high fouling with the same membrane material. The Clackamas river water had low TOC, high SUVA, and high silica compared to the Chattahoochee River water. Comparing the flux decline curves and water quality from these two sources does not lead to a definitive conclusion with regard to water quality factors that may contribute to fouling of PVdF membranes. The Waikato and Chattahoochee Rivers caused similar amounts of fouling but the water quality of the two rivers does not appear similar.

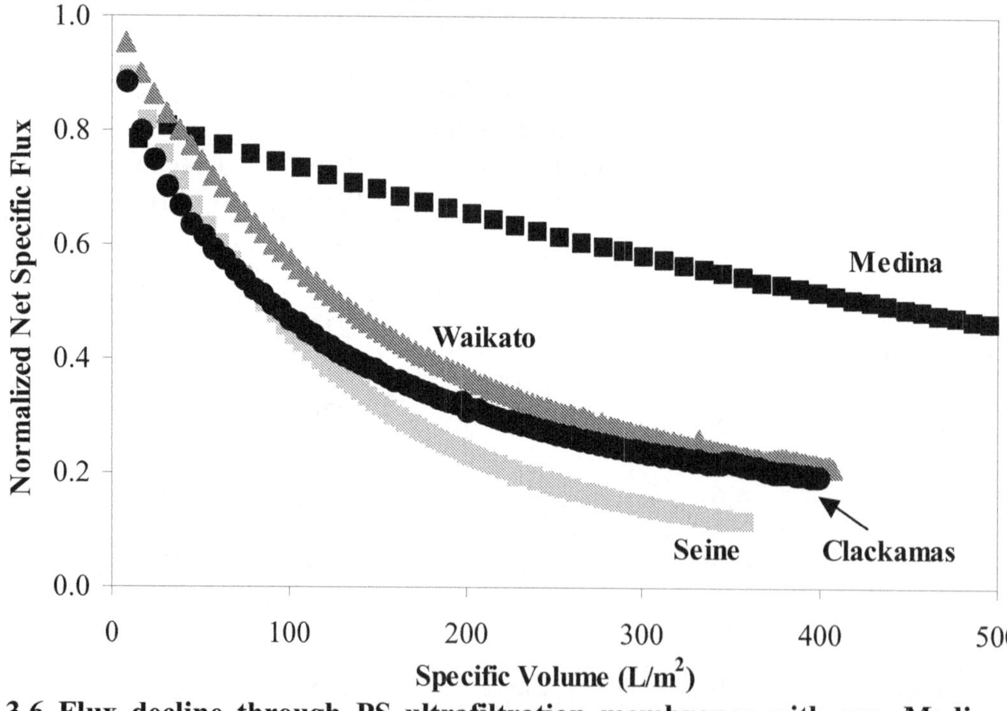

Figure 3.6 Flux decline through PS ultrafiltration membranes with raw Medina, Seine, Clackamas, and Waikato River waters

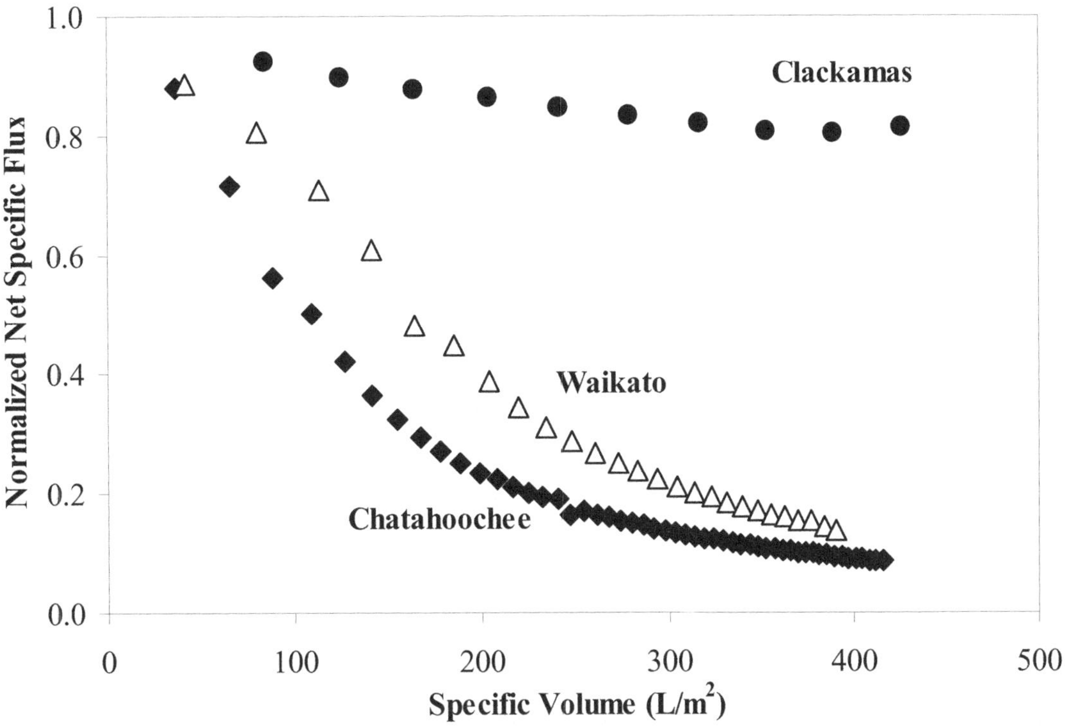

Figure 3.7 Flux decline through PVdF microfiltration membranes with raw Chattahoochee, Clackamas, and Waikato River waters

A difference between membrane materials can be observed by comparing of Figure 3.6 and Figure 3.7. The greater fouling of Clackamas River raw water through the PS ultrafiltration membrane when compared to the PVdF microfiltration membrane is shown in Figure 3.8. The Clackamas River is a low TOC, low alkalinity (5-10 mg/L as $CaCO_3$), and relatively high silica concentration water. Despite having a low TOC concentration, the Clackamas River has a relatively high UV_{254} absorbance compared to some of the other source waters, leading to a high SUVA value. The high silica and SUVA values do not appear to cause significant fouling of PVdF microfiltration membrane. However, the PS ultrafiltration membrane fouled significantly with Clackamas River water even though the water had a low TOC concentration, which points out that TOC is not the only factor responsible for fouling of PS ultrafiltration membranes. The combination of high silica and SUVA values may be one possible factor that contributed to the fouling of the PS membrane with this water. These experiments suggest that the PVdF membrane may be more suitable than the PS membrane for treating raw Clackamas River water.

In contrast to the Clackamas River water, the Waikato River water cause significant fouling of both PS and PVdF membranes, as shown in Figure 3.9. The difference in behavior of the Clackamas and Waikato Rivers when filtered through similar membranes demonstrates that the membrane material and the source water quality are both important factors in influencing the rate of membrane fouling.

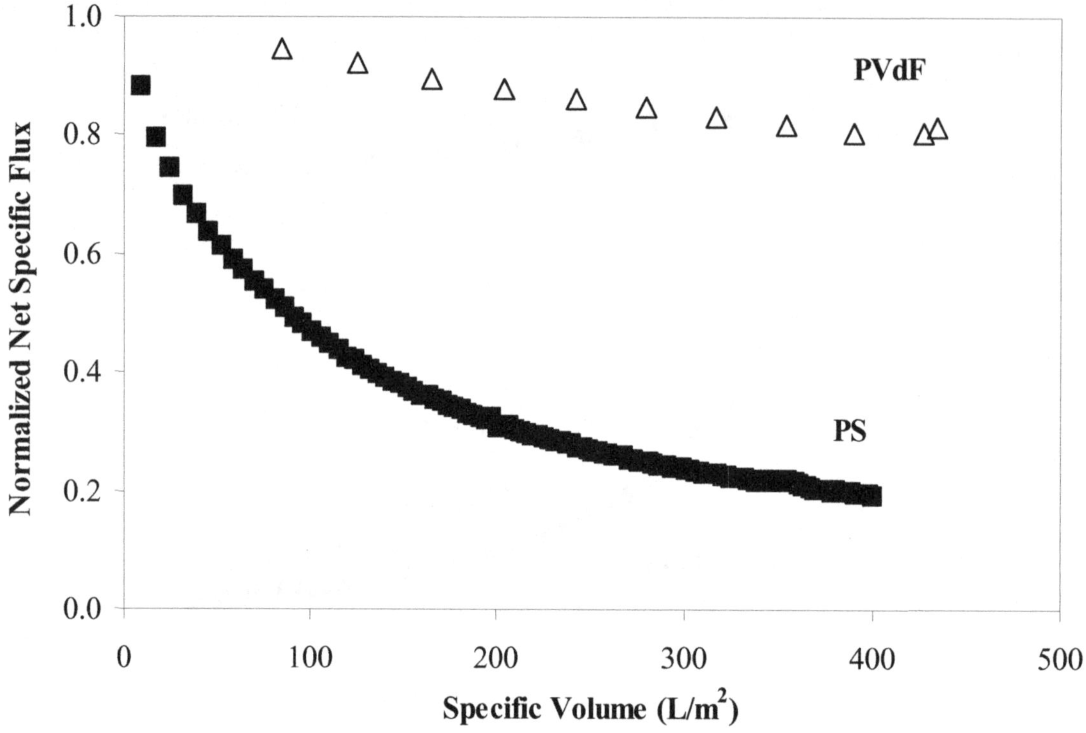

Figure 3.8 Flux decline of raw Clackamas River water through PS and PVdF membranes

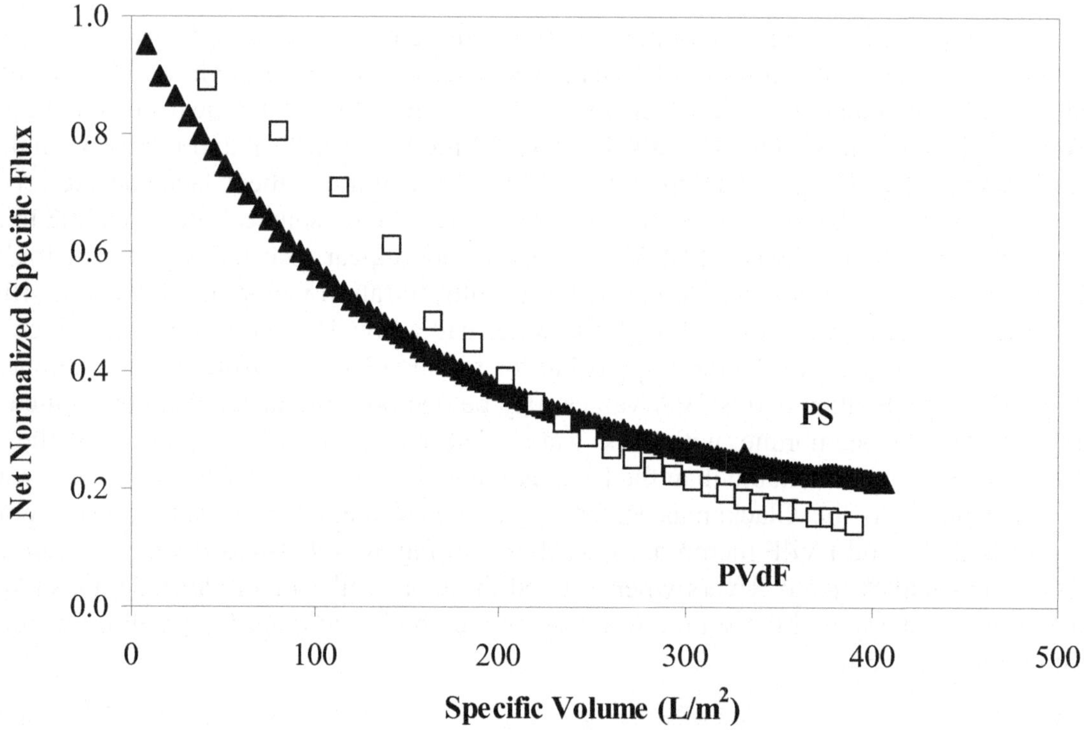

Figure 3.9 Flux decline of raw Waikato River water through PS and PVdF membranes

SEM Characterization of Clean and Fouled Membranes

Scanning electron microscopy can provide morphological information on fouling observed with various source waters with the same membrane material. An SEM image of an unused PVdF membrane is shown in Figure 3.10. Figure 3.11 and Figure 3.12 give visual indication of fouling on PVdF membranes with raw Waikato and Chattahoochee River waters. The deposition on the surface of the Waikato River membrane appears to be a cake-like structure and the deposition on the surface of the Chattahoochee River membrane appears to be more agglomerated particulates in nature. In both fouled membrane images, however, the membrane surface is completely covered with foulants and the clean membrane structure (Figure 3.10) is not visible, suggesting a uniform deposition of fouling material. In contrast, a different degree of fouling is suggested by the image of a PVdF membrane fouled by raw Clackamas River water in Figure 3.13. Although material from the raw water is deposited on the membrane, the original membrane structure is still clearly visible, suggesting that the deposition on the membrane from Clackamas River water is much thinner than the deposition from the other two waters. These observations support the flux analysis in Figure 3.7 in which both the Chattahoochee River and Waikato River waters have greater fouling than the Clackamas River.

An SEM image of a PS membrane fouled with raw Waikato River water is shown in Figure 3.14. Figure 3.9 demonstrated that the PS and PVdF membranes had a similar extent of fouling after filtering Waikato River water. A comparison of the SEM images in Figure 3.11 and Figure 3.14 show similar cake layer structure formation on the surface of the membrane, which may explain the similarities in membrane performance.

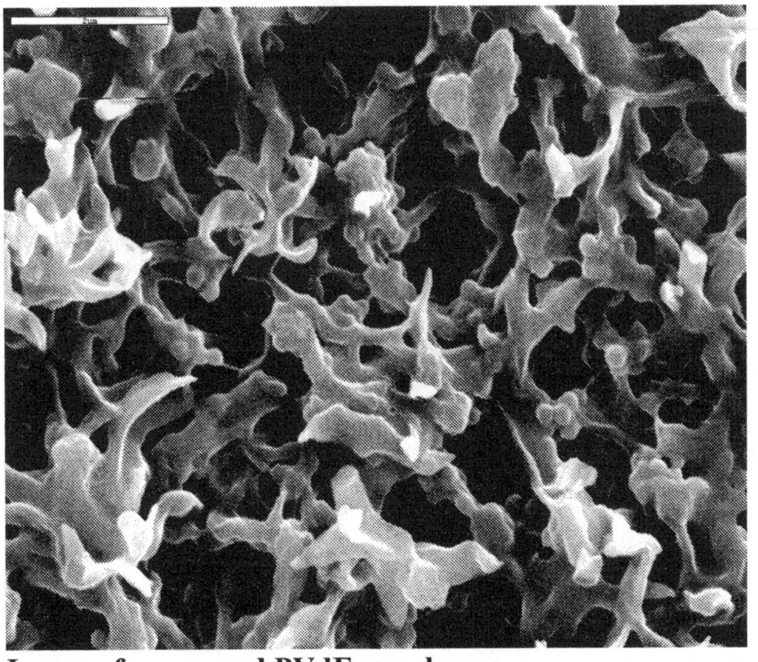

Figure 3.10 SEM Image of an unused PVdF membrane

Figure 3.11 SEM image of a PVdF membrane fouled with raw Waikato River water

Figure 3.12 SEM image of a PVdF membrane fouled with raw Chattahoochee River water

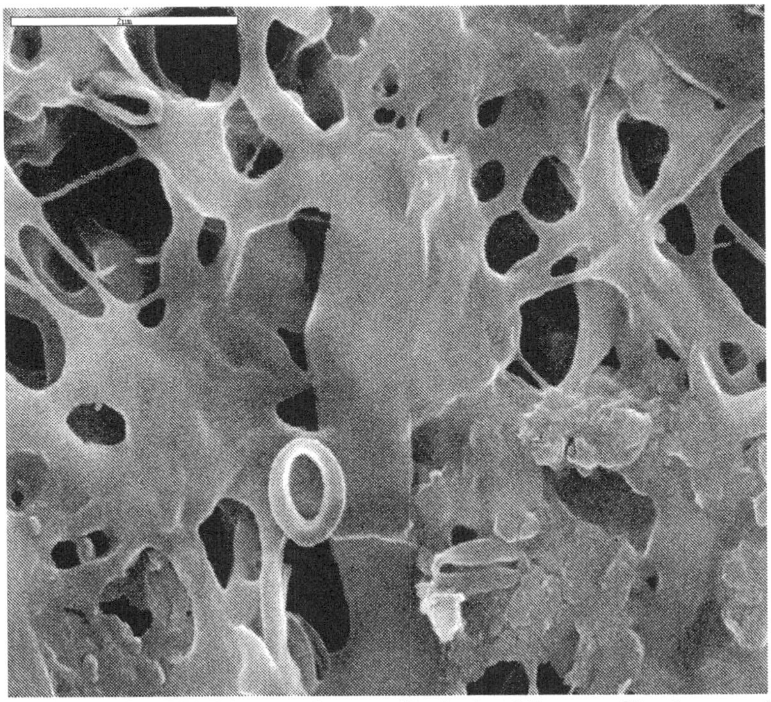

Figure 3.13 SEM image of a PVdF membrane fouled with raw Clackamas River water

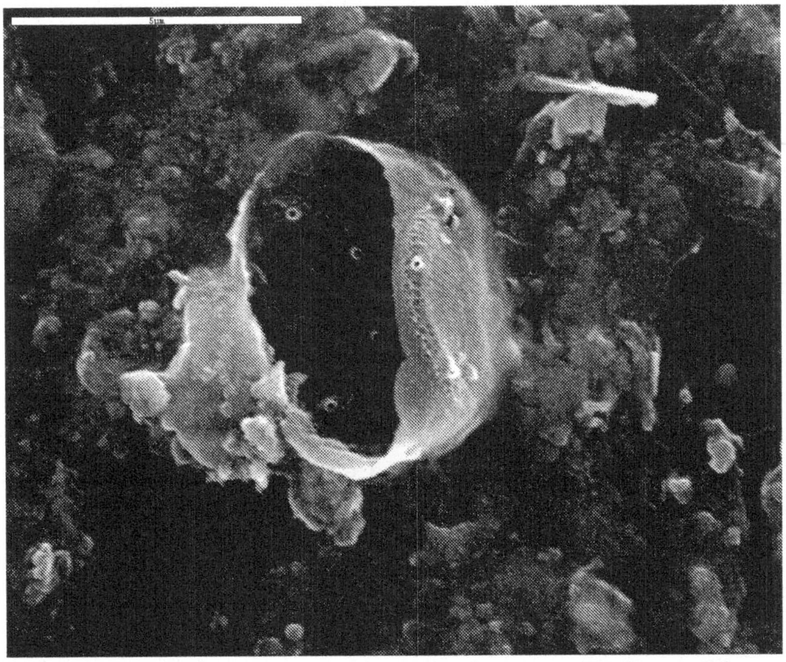

Figure 3.14 SEM image of a PS membrane fouled with raw Waikato River water

FTIR Spectrometry of Fouled Membrane Surfaces

FTIR spectrometry was used to examine the functional groups of the foulants. A clean PVdF membrane and a PVdF membrane fouled with Waikato River water are compared in Figure 3.15. The spectra in the figure demonstrate that the surface chemistry changed due to fouling of the PVdF membrane. The peaks observed in the clean membrane spectrum are no longer visible when compared to the fouled membrane spectrum, which could be due to masking of the original surface by the foulant layer on the membrane. The primary peaks in the Waikato River fouled membrane spectrum are a broad band of absorbance between 3700 and 2900 cm^{-1}, which is indicative of the hydroxyl functional group, a smaller peak at 1645 cm^{-1}, which is indicative of amine functional groups, and a large peak at 1038 cm^{-1}, which is typically due to either C-O bonds or Si-O bonds. The absorbance suggests fouling by either organic matter and/or silica.

FTIR spectrometry can also provide insight into the different fouling characteristics associated with different water and membrane materials. A comparison of the FTIR spectra of PS and PVdF membrane fouled with raw Clackamas River water is shown in Figure 3.16 and Figure 3.17. The spectrum of the fouled PS membrane shows a significant peak at 1645 cm^{-1} that is not present in the spectrum of the fouled PVdF membrane with the same water (Figure 3.17). This peak is often attributed to amines, which could signify the presence of NOM. This observation indicates that PS membrane is not suitable for raw water with high levels of SUVA such as Clackamas River water. As shown in Figure 3.8, the fouling of this source water with these two membrane materials was significantly different. These results suggest that a particular constituent in the feed water may interact with one type of membrane material but not a different membrane material.

A similar comparison can be performed to examine the differing nature of two source waters filtered through the same type of membrane material. The FTIR spectra from PVdF membranes fouled with raw Waikato River water is shown in Figure 3.18. As shown in Figure 3.17, the Clackamas River water spectrum has a number of pronounced peaks that look similar to the FTIR spectrum of a clean PVdF membrane (Figure 3.15), suggesting that there was little deposition on the membrane, and the absorbance due to the base material was evident within the spectrum. The SEM images of these fouled membranes were shown in Figure 3.11 and Figure 3.13. The extent of material on the membrane observed in the SEM images were consistent with these observations from the FTIR spectra. The Waikato River fouled PVdF membrane spectrum is considerably different and the peaks associated with the PVdF material have been masked by the deposition of material from the feed water. The greater fouling shown in Figure 3.7 supported the observation from FTIR spectra.

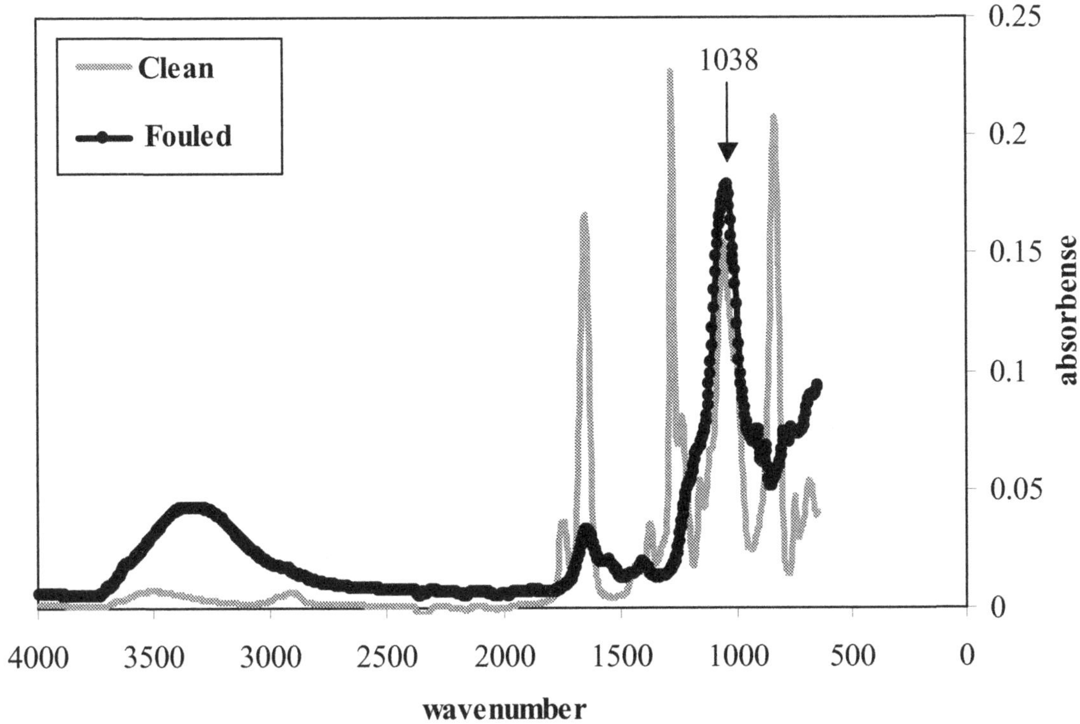

Figure 3.15 FTIR spectra of clean and Waikato-fouled PVdF membranes

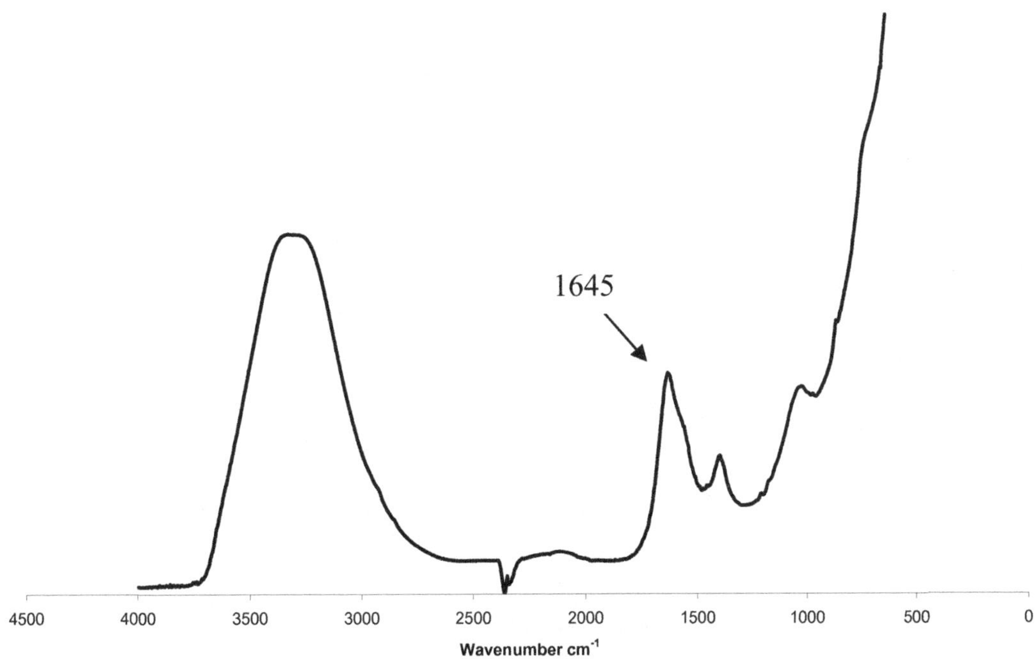

Figure 3.16 FTIR spectra of PS membrane fouled with raw Clackamas River water

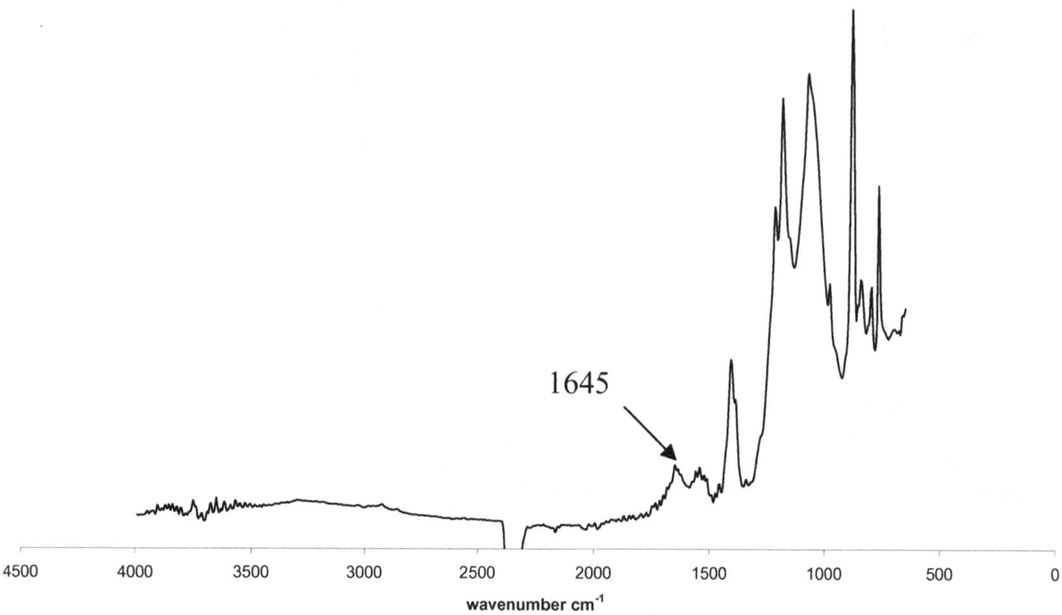

Figure 3.17 FTIR spectra of PVdF membrane fouled with raw Clackamas River water

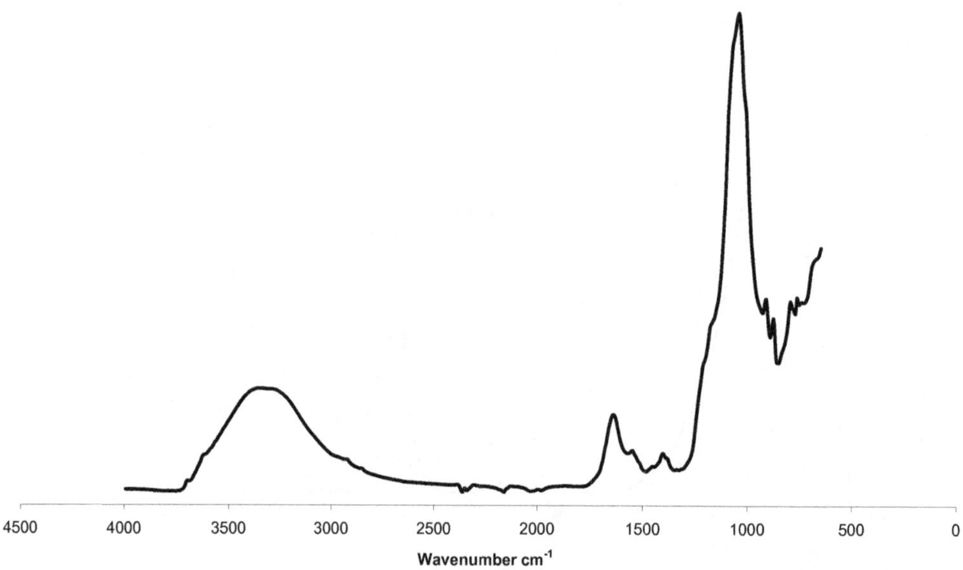

Figure 3.18 FTIR spectra of PVdF membrane fouled with Waikato River water

Summary of the Effect of Source Water on Membrane Flux

The effects of the source water selection on membrane performance can be summarized as follows:

1. No single measured parameter, such as pH, alkalinity, hardness, TOC, UV254 absorbance, SUVA, or silica could be correlated to the extent of fouling. A combination of several factors, or water quality characteristics that are not adequately described by current parameters, may be responsible for membrane fouling.
2. Source waters interacted with different membrane materials in different ways. The Clackamas River water caused minimal fouling of the PVdF membrane and extensive fouling of the PS membrane, whereas the Waikato River caused extensive fouling of both membrane materials.
3. SEM provided useful insight into the fouling caused by natural waters. SEM images of PVdF membranes fouled with Chattahoochee, Waikato, and Clackamas River waters showed extensive fouling by the Waikato and Chattahoochee Rivers and minimal fouling by the Clackamas River, which was consistent with the flux results.
4. FTIR spectra of fouled membranes of different materials or different source waters showed different surface chemistry. In a qualitative sense, some differences could be related to differences in membrane performance. For instance, the FTIR spectrum of Clackamas River water showed several peaks associated with the base PVdF membrane material and the FTIR spectrum of the Waikato River water filtered through the same membrane did not show the similar peaks. This observation suggested that the foulant layer on the membrane with the Clackamas River water was thinner than it was on the membrane with Waikato River water, which was supported by both the flux data and the SEM images. Overall, however, the FTIR did not provide useful insight into the specific nature of the foulants on various membrane materials when exposed to different source waters.

Effect of Pretreatment on Membrane Flux

A significant component of the bench scale testing of this study was an evaluation of the effect of pretreatment chemicals such as coagulants on membrane performance. A comparison of the performance of PS ultrafiltration membranes during filtration of raw and clarified Medina, Seine, and Clackamas River waters is shown in Figure 3.19. As was shown in Figure 3.6, the Medina River water showed the least fouling, with greater fouling in the Seine and Clackamas Rivers. In all cases, coagulation and clarification improved membrane performance. The improvement in flux in Seine River water was more than in the Medina River and Clackamas River water.

The coagulant doses used for the bench scale studies were doses optimized at the utility. The effect of coagulation and clarification on water quality is shown in Table 3.9. TOC removal during coagulation and clarification ranged from 12 to 42 percent in the five source waters, and silica removal ranged from 0 to 20 percent. Of the three source waters shown in Figure 3.19, the greatest TOC removal was observed in the Seine River water, which may relate to the greatest improvement in membrane performance.

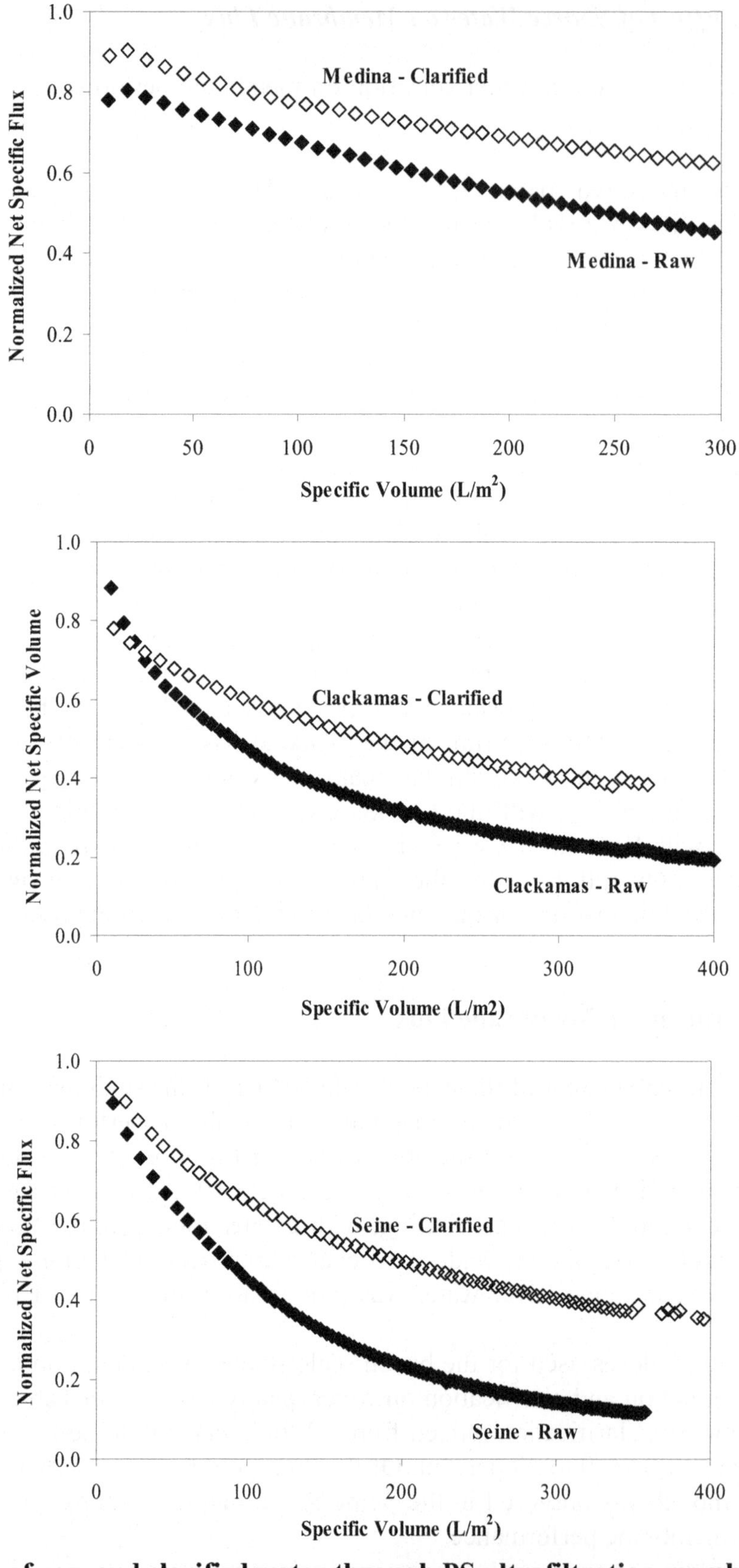

Figure 3.19 Flux of raw and clarified water through PS ultrafiltration membranes

Table 3.9
Clarified water quality data

Parameter	Medina River	Chattahoochee River	Seine River	Clackamas River	Waikato River
TOC (mg/L)	1.19	0.97	2.24	0.76	1.11
UV_{254} (cm^{-1})	0.042	0.032	0.054	0.018	0.036
SUVA (L/mg·m)	3.5	3.3	2.4	2.4	3.3
Silica (mg/L)	13.9	6.9	2.88	18.8	26.8
TOC percent removal	12	21	26	15	42
Silica percent removal	10	—	—	5	20

A comparison of the flux of raw and clarified Chattahoochee River, Waikato River, and Clackamas River water through the PVdF microfiltration membranes is shown in Figure 3.20. The change in flux with PVdF membranes after coagulation and clarification is different from the change in flux with PS membranes.

The raw Chattahoochee River water showed high fouling but the fouling decreased when the water was clarified. This suggested that coagulation could enhance membrane performance in Chattahoochee River water. The raw Clackamas River water had very little fouling when passed through PVdF membrane. After coagulation and clarification, however, the flux decreased significantly. This result was noteworthy since the measured water quality parameters (such as TOC, UV_{254}, and silica concentration) decreased after clarification of Clackamas River water, but the fouling increased. The decrease in flux was opposite to what was observed in Figure 3.19, where the PS ultrafiltration membrane performance increased after clarification of Clackamas river source water. It should also be noted that the performance of PS and PVdF membranes was substantially different with raw Clackamas River water (recall Figure 3.8). These results suggest that the constituents in Clackamas River water that contribute to fouling of PS ultrafiltration membranes are different from those that contribute to fouling of PVdF microfiltration membranes, and that the response of these different constituents to coagulation and clarification is also different.

The results from the Waikato River fell between the results for the Chattahoochee and Clackamas Rivers. Early in the filter run, the fouling after clarification was less than without clarification, but by the end of the filter run, the fouling with clarification was slightly worse than without clarification. One similarity between the Clackamas and Waikato River waters was that both had relatively high silica concentrations. One possible explanation for the lack of improvement in flux after clarification may be the formation of a fouling species due to interactions between the coagulant and silica.

In summary, coagulation and clarification was beneficial to the performance of PS ultrafiltration membranes in all cases in this study, but the results were more site-specific with the PVdF microfiltration membranes, leading to an improvement in performance in one case, little change in a second case, and worse performance in the third case.

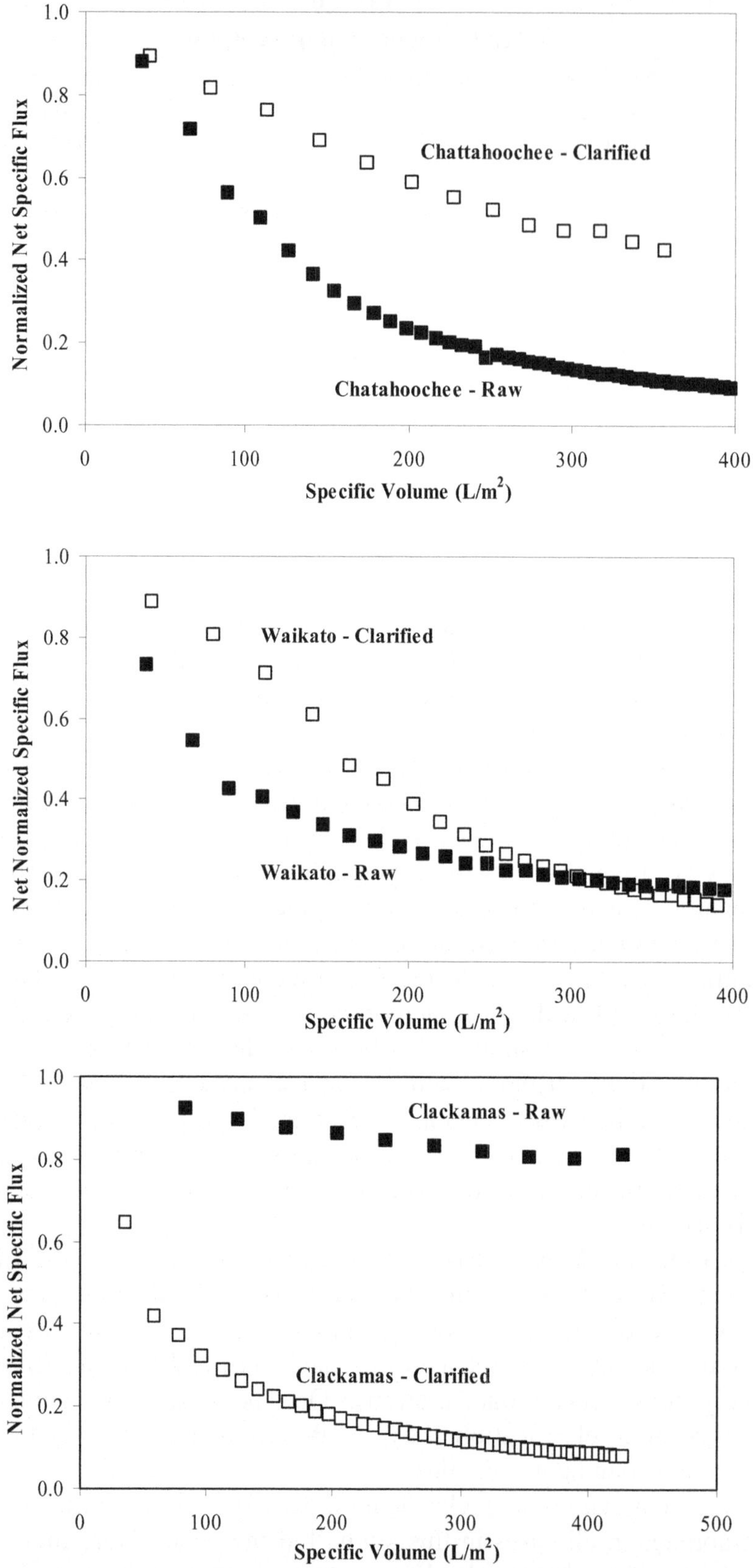

Figure 3.20 Flux of raw and clarified water through PVdF microfiltration membranes

Comparison of Bench Clarification to Pilot Clarification for Bench Membrane Flux Tests

In a small number of studies, water clarified at the location of the pilot- or full-scale plant was compared to water clarified in the laboratory. This was accomplished by collecting a raw water sample and a coagulated or clarified water sample from the utility, and shipping each water sample to UNM. The raw water sample was then used to perform coagulation in the laboratory. The effect of the location of coagulation on membrane performance using the Medina River water is shown in Figure 3.21. For this source water, feed water was collected from the raw water and from the effluent from the Superpulsator (clarifier) at the full-scale plant, representing untreated water, and water that had already been treated by coagulation and clarification. A comparison of the water quality of each of these waters is shown in Table 3.10. Although most water quality parameters are similar, the Superpulsator effluent has lower TOC and lower SUVA than the laboratory-coagulated water. These differences may have resulted in differences in membrane performance. In Figure 3.21, the coagulation in the laboratory resulted in an improvement in flux, but the filtration of the Superpulsator effluent showed a decrease from the raw water flux performance. Ferric sulfate was used in the laboratory and ferric chloride was being used at the plant, but the lab coagulation used the same dose of metal ion (6.87 mg/L as Fe) as what was being used at the full-scale facility. It should be noted that the Medina River water was the first source water tested during this research, and an excessive amount of time might have passed between the collection of the Superpulsator effluent and the membrane flux tests in the laboratory. Nevertheless, the results suggest that membrane flux tests using clarified water from a utility may not produce results that are consistent with coagulation and membrane filtration studies done in a laboratory.

Similar tests were carried out using Chattahoochee River water and PVdF membranes, as shown in Figure 3.22. In this group of tests, three feed waters were collected from the Atlanta Fulton County Water Treatment Plant (AFCWTP) pilot plant, representing untreated water, coagulated but not clarified water (flocculated), and coagulated/clarified water. The results of coagulation and clarification at the pilot plant can be compared with settled water in the laboratory in Figure 3.22. With this source water, coagulation without clarification caused a quicker flux decrease in the beginning and resulted in slight better performance than the raw water, but the addition of clarification caused improved performance in both laboratory-clarified water and pilot-clarified water. However, the laboratory and pilot coagulated waters again show significantly different flux curves, which is consistent with the result from the Medina River water, that using clarified water from a utility may not produce results that are consistent with laboratory coagulation and membrane filtration studies.

As will be noted later in this report, coagulation and membrane filtration studies carried out in the laboratory typically agreed qualitatively with pilot studies, suggesting that bench-scale studies may be best conducted by performing the coagulation and clarification in the laboratory, with more controlled conditions, rather than shipping clarified water from a utility. It is possible that the long time that passes between coagulation and membrane filtration (due to the shipping time) causes changes in water quality that are not adequately reflected in the membrane studies.

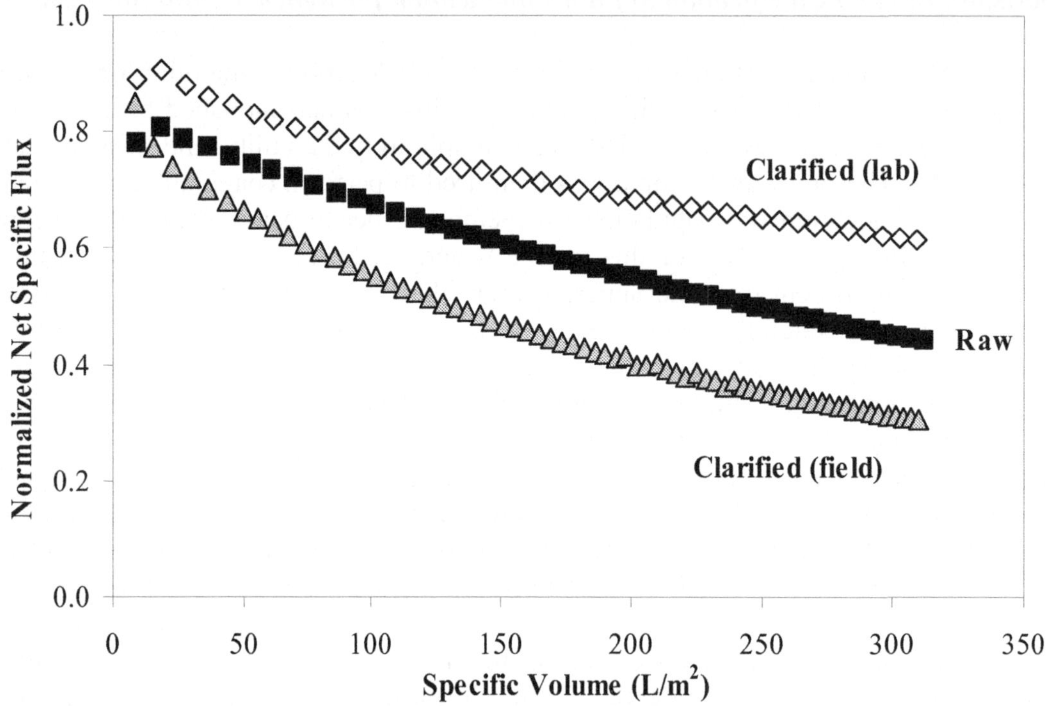

Figure 3.21 Flux decline of laboratory and full-scale clarified Median River water through PS microfiltration membranes

Table 3.10
Medina River raw, laboratory-clarified, and Superpulsator effluent water quality

Parameter	Raw water	Coagulated in lab	Superpulsator
pH	8.4	8.3	8.3
Alkalinity (mg/L as $CaCO_3$)	219	200	185
Hardness (mg/L as $CaCO_3$)	248	248	248
Calcium (mg/L)	66	67	66
Magnesium (mg/L)	20.2	19.5	20.2
Iron (mg/L)	0.0098	0.064	0.006
Silica (mg/L)	15.4	13.9	12.4
TOC (mg/L)	1.35	1.19	1.07
UV_{254} (cm^{-1})	0.054	0.042	0.033
SUVA (L/mg·m)	3.98	3.56	3.08

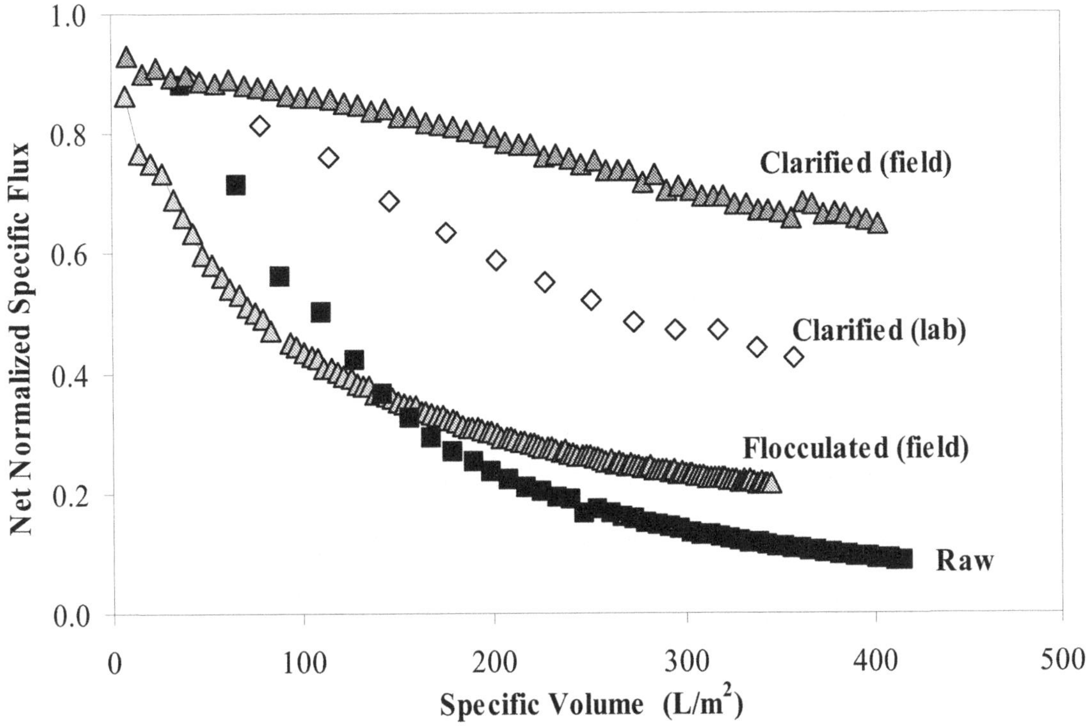

Figure 3.22 Comparison of flux decline of laboratory and pilot-scale flocculated and clarified Chattahoochee River water for PVdF membranes

SEM Characterization of Membranes Fouled with Clarified Water

Scanning electron microscopy was used to evaluate the physical change in the fouling layer following coagulation and clarification. The SEM image in Figure 3.12 gives a picture of extensive fouling with raw Chattahoochee River water through the PVdF membrane. The extensive fouling from raw water was also evident form the flux curve in Figure 3.20. After coagulation and clarification with a low alum dose of 7 mg/L, the flux decline rate decreased. The SEM image in Figure 3.23 shows a cake layer formation that could be porous as increase in membrane flux was observed after clarification.

The SEM image in Figure 3.13 shows a PVdF membrane fouled with raw Clackamas River water. As noted earlier, the surface in this image is considerably cleaner than the surfaces of other membranes after filtration of other raw surface waters. This observation was consistent with the flux results. With Clackamas River water, a substantial increase in fouling was observed after coagulation and clarification. An SEM image (Figure 3.24) of a PVdF membrane fouled with clarified Clackamas River water supports this increase in fouling. After clarification, the surface of the PVdF membrane is completely coated with a cake layer. This material could be due to an interaction between the coagulant and a component in the raw water, such as silica. The material shown in Figure 3.24 could contribute to the extensive fouling observed in the flux results in Figure 3.20.

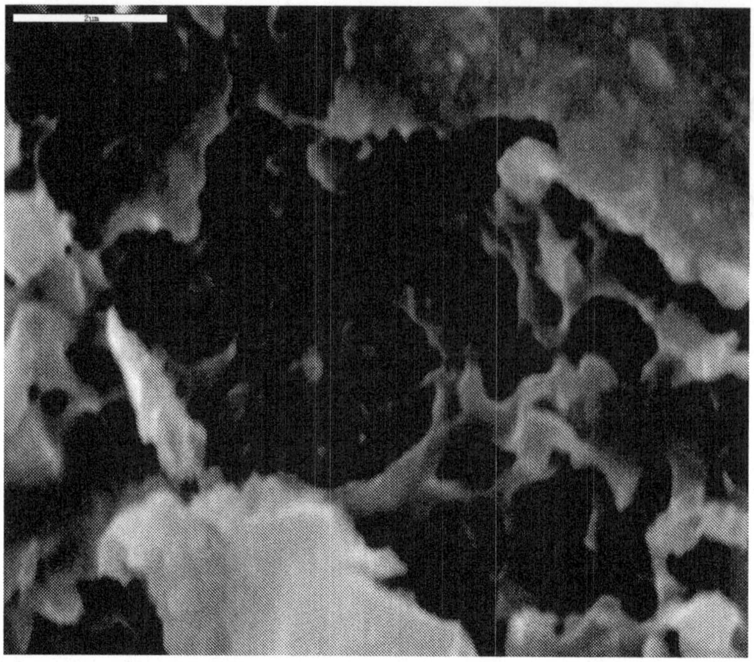

Figure 3.23 SEM image of a PVdF membrane fouled with alum-coagulated and clarified Chattahoochee River water

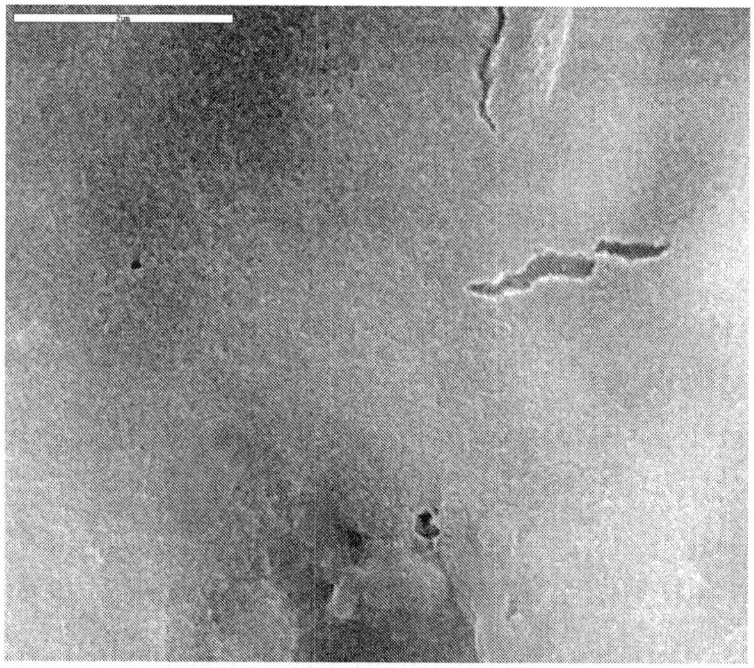

Figure 3.24 SEM image of a PVdF membrane fouled with alum-coagulated and clarified Clackamas River water

FTIR Characterization of Membranes Fouled with Clarified Water

FTIR spectrometry was also used to elucidate the changes in fouling following clarification. A comparison of FTIR spectra of membranes fouled with raw and clarified

Clackamas River waters is shown in Figure 3.25. The spectrum of the membrane fouled with raw water, as noted earlier, shows a number of peaks associated with the PVdF material. The peaks are obscured in the spectrum of the membrane fouled with clarified water. The primary peaks in the clarified Clackamas River spectrum are twin peaks at 1635 and 1519 cm^{-1}, which are indicative of amine functional groups, and a large peak at about 1000 cm^{-1}, which is typically due to either C-O bonds or Si-O bonds. The absorbance suggests that the fouling is caused by either organic matter or silica, which is also consistent with the figures of flux decline that show higher fouling after clarification of Clackamas River water. It should be noted that the major peaks in clarified Clackamas River water were observed earlier on raw Waikato River water fouled PVdF membrane (Figure 3.15).

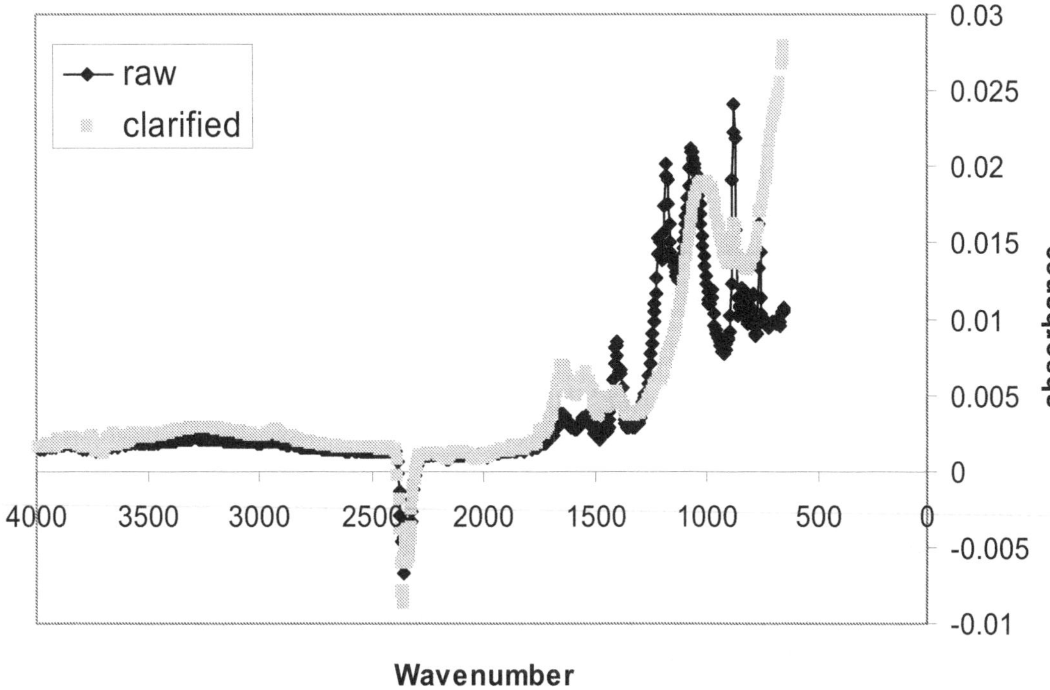

Figure 3.25 FTIR spectra of PVdF membranes fouled with raw and clarified Clackamas River water

Effect of Alternate Pretreatment Chemicals on Flux Decline

Pretreatment chemicals other than coagulants including PAC and chlorine were also in bench-scale study. In addition, multiple coagulants were tested with the same source water in two cases. Flux through PS ultrafiltration membranes following pretreatment of Seine River water with alum, ferric chloride, and PAC is shown in Figure 3.26. The most extensive fouling curves in the figure are raw Seine river water and the PAC-treated raw water, which suggests that a low dose of PAC (2 mg/L) does not improve membrane performance. Alum and ferric chloride coagulation used doses provided by the participating utility, which, based on the reported concentrations, provides a similar molar dose (1.27×10^{-4} mol/L for Fe and 1.17×10^{-4} mol/L for Al). Based on the similarity of doses, these data suggest that alum is a better coagulant for improving membrane flux than ferric for this particular source water.

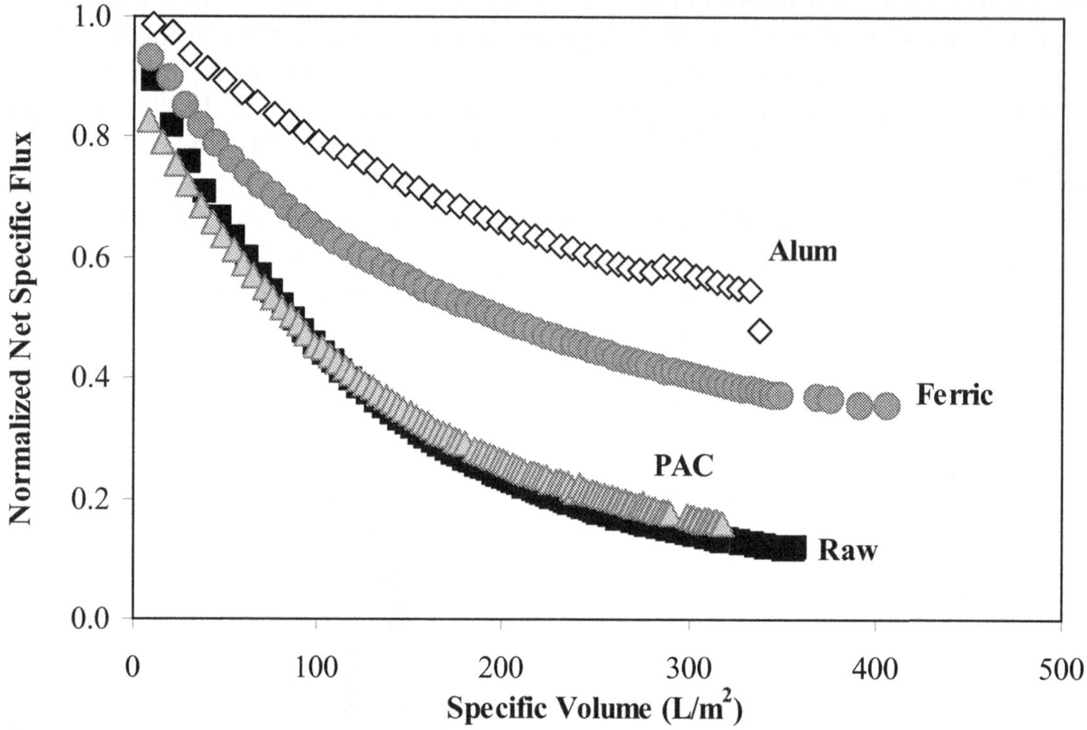

Figure 3.26 Flux of Seine River water after pretreatment with alum, ferric sulfate, and PAC

Multiple pretreatment chemicals were tested with Clackamas River water. The flux decline of Clackamas River water through PVdF membranes after pretreatment with alum, ferric sulfate, ACH, and chlorine is shown in Figure 3.27. The same dose of chemical was calculated and used for clarification. The least fouling occurred with untreated raw water, and chlorine-pretreated raw water. Chlorine did not change the flux for the Clackamas River water in these bench scale studies. As discussed in the pilot-scale section, it is suspected that chorine controls biofouling that would normally shows the effect when membranes are operated for long term.

Alum and ferric chloride pretreatment fouled the membrane extensively. As noted earlier, coagulation and clarification made fouling worse in this water, which may be due to the formation of a foulant caused by the interaction of the coagulant with the silica in the water. Although all coagulants made the fouling worse, ACH was the comparatively best coagulant that was tested with Clackamas River water. Higher doses of alum (10 mg/L, not shown) helped in improving flux but did not substantially bring it to the initial raw water flux.

The fouling of Clackamas River water through PS membranes after pretreatment with ferric sulfate, alum, and ACH is shown in Figure 3.28. The greatest fouling occurred with raw water. As noted earlier (Figure 3.8), the fouling of PS membranes is substantially different from what was observed in Figure 3.27, where raw Clackamas River water fouling on PVdF membrane was very minimal. After coagulation and clarification, however, fouling decreased in PS ultrafiltration membrane. Similar to the results with PVdF membrane, the extent of fouling with alum and ferric chloride clarified water was almost the same. Coagulation and clarification with ACH produced the least amount of fouling among the tested coagulants with both the PS and PVdF membranes.

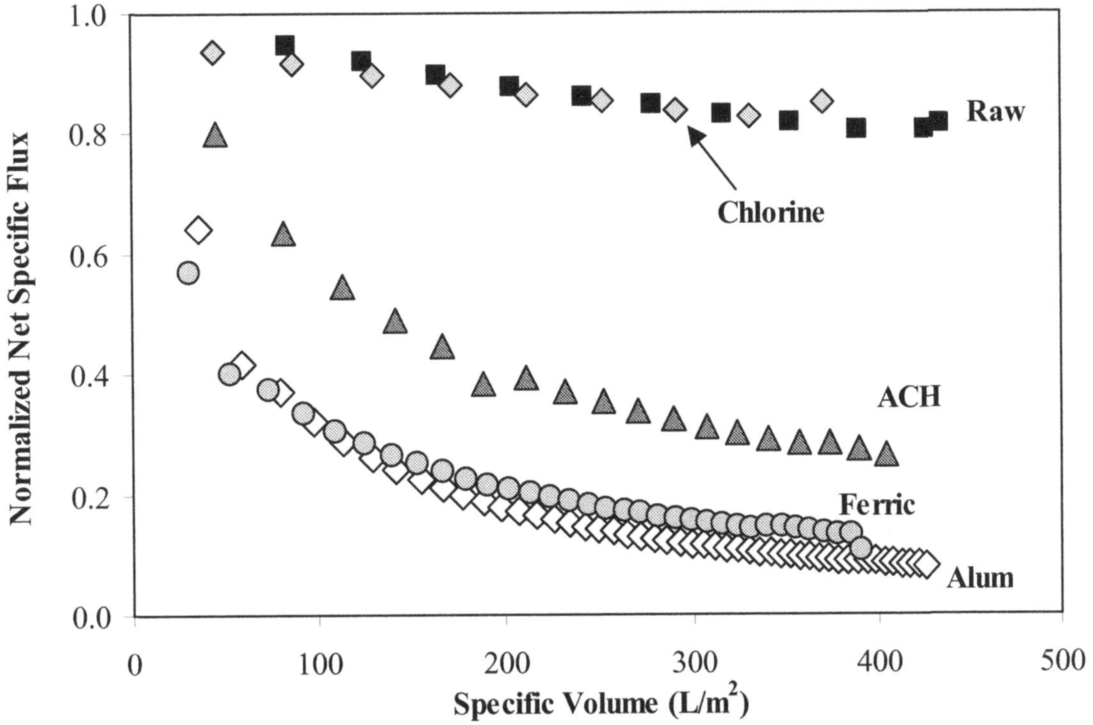

Figure 3.27 Fouling of Clackamas River water pretreated with various chemicals through PVdF membranes

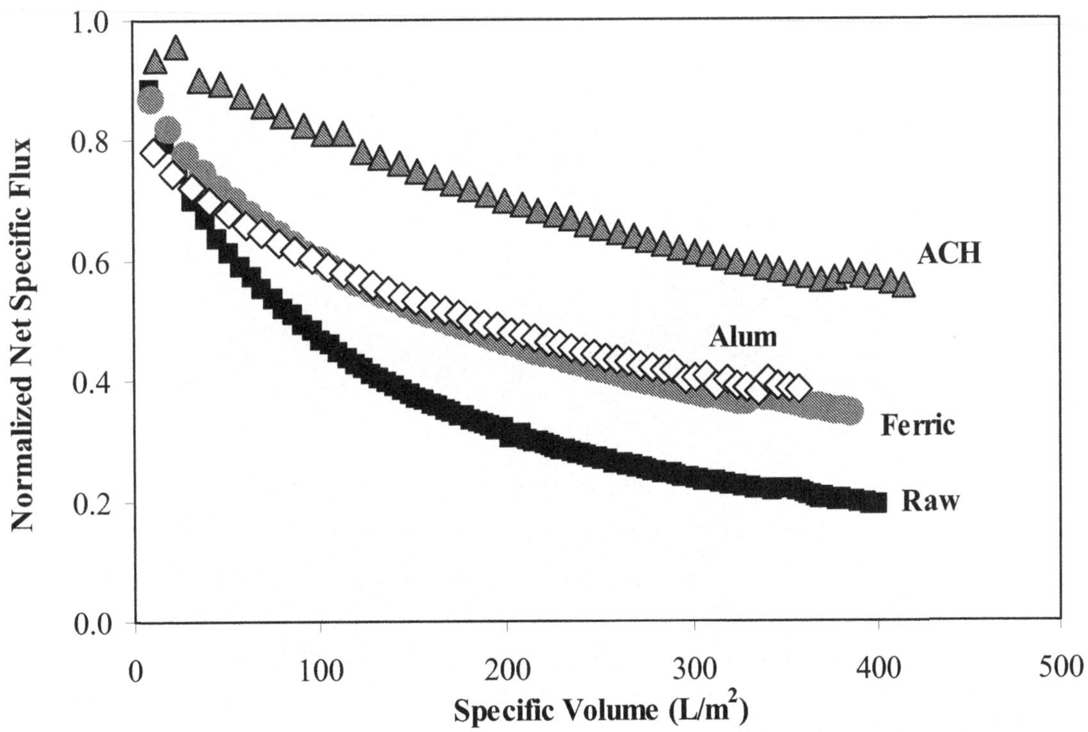

Figure 3.28 Fouling of Clackamas River water pretreated with various chemicals through PS membranes

SEM images of PVdF membranes fouled with Clackamas River water clarified with various coagulants are shown in Figure 3.24, Figure 3.29, and Figure 3.30. In each case, the membrane is coated with a cake of material that obscures the membrane, suggesting a physical reason for the increase in fouling compared to raw water (Figure 3.13). However, the lower fouling from ACH-clarified water is not evident in the SEM images, indicating that SEM can only provide qualitative indications of the degree of fouling of membranes.

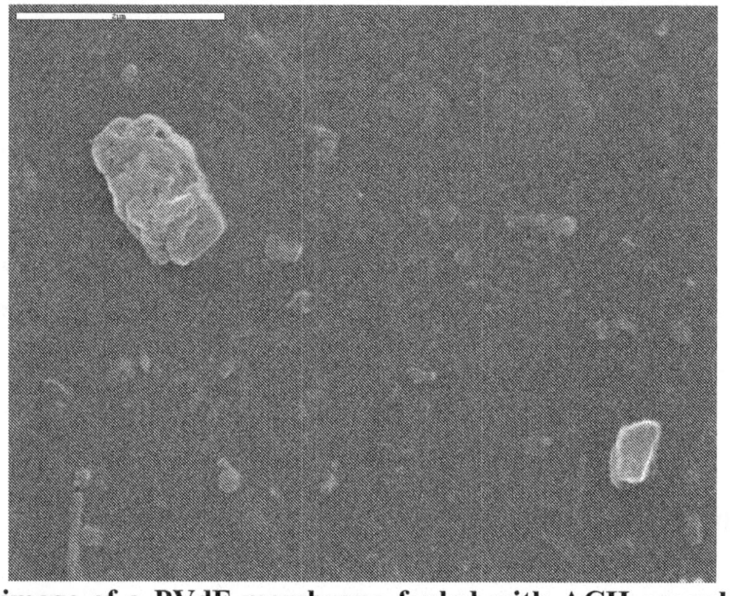

Figure 3.29 SEM image of a PVdF membrane fouled with ACH-coagulated and clarified Clackamas River water

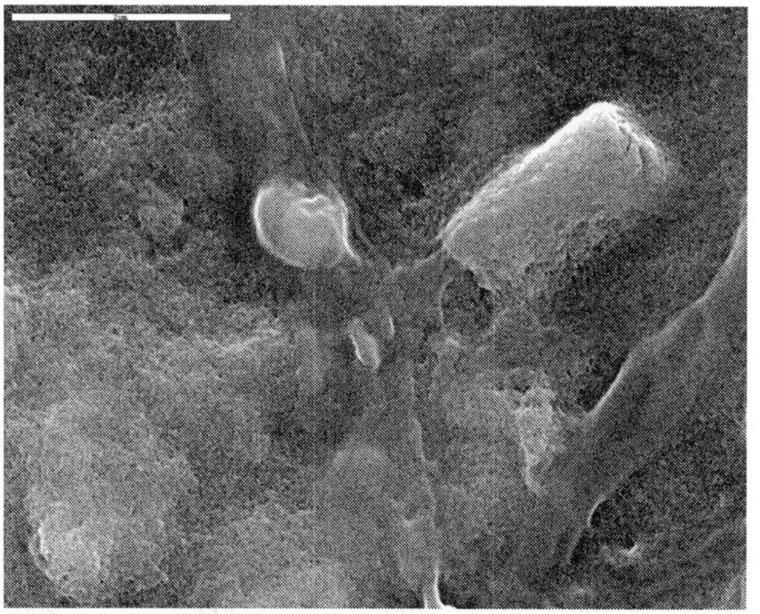

Figure 3.30 SEM image of a PVdF membrane fouled with ferric sulfate-coagulated and clarified Clackamas River water

A comparison of the fouling of PS and PVdF membrane with Clackamas River water pretreated with a low chlorine dose is shown in Figure 3.31. The PVdF membrane experienced very little fouling with raw Clackamas River water, and chlorine had no effect on the rate of fouling. In contrast, the PS membrane had significant fouling with raw Clackamas River water, and had worse fouling after 1 mg/L of chlorine was added to the raw water. This observation can be attributed to the fact that chlorine might be responsible for causing changes in the NOM character of the raw water, which had an impact on PS membranes but not on PVdF membranes. Further research is required to investigate the impact of chlorine on membrane foulants.

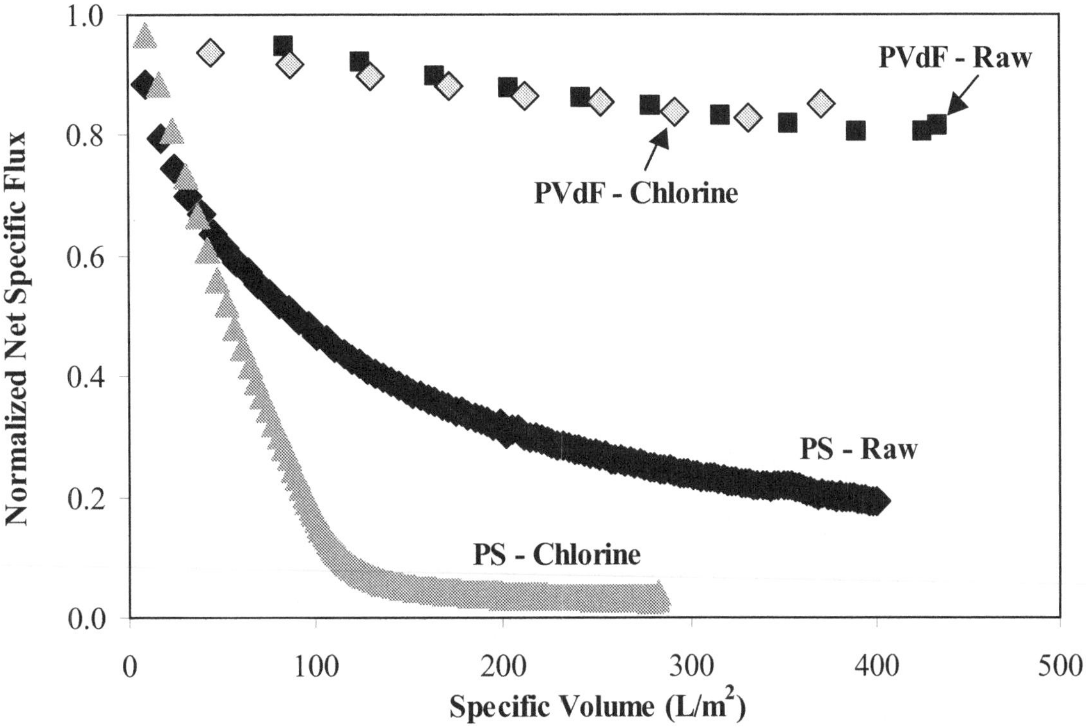

Figure 3.31 Comparison of fouling of PS and PVdF membrane with chlorine-treated Clackamas River water

Comparison to Membrane Performance Results from a Previous AwwaRF Study

One of the source waters in this study, the Medina River, was also used in a previous AwwaRF study, *Coagulation Pretreatment for Membrane Filtration* (Howe and Clark 2002a). As a result, data from the previous study can be used to support and extend some of the data collected during this study. One early objective of this study was to evaluate the CA membrane at both bench and full scale using the Medina River water. Unfortunately, the flat sheet CA membrane received for the bench scale studies had manufacturing difficulties and could not produce a reasonable permeate flux. As a result, the CA membrane was not used at the bench scale, even though it is the membrane installed in the plant that uses the Medina River as a water source.

The majority of pilot-scale tests conducted at Centre International de Recherche sur l'Eau et l'Environment (CIRSEE) (see pilot-scale section for details) used CA membranes. For comparing bench- and pilot-scale studies, data from the previous study is included here to provide a comparison between the CA and PS membranes. A comparison between raw and coagulated water using the CA and PS membranes, and using the previous study and the current data, is shown in Figure 3.32.

The flux of uncoagulated water with the PS membrane is nearly identical between the previous and current studies, demonstrating the usefulness of using data from the previous study. In both studies, addition of a coagulant improved membrane flux. With this raw water, the CA membrane performed better than the PS membrane. For all coagulant tested, clarification improved membrane performance. The same dose of alum (4.5 mg/L as Al), used with the CA and PS membranes in the previous study resulted in similar improvements on membrane flux. The flux increased from 60 percent of the initial flux for raw water to 76 percent of the initial flux for coagulated water for the CA membrane, and increased from 52 percent to 66 percent of the initial flux after coagulation with the PS membrane. The increase in flux was slightly lower with the PS membrane in the current study, but the data was collected using a different coagulant and a different dose.

The effect of different coagulants with the Medina River water and the CA membrane in the previous study is shown in Figure 3.33. These results show that alum and ferric sulfate produced similar flux through the CA membrane when used at the same molar dose, but PACl had lower flux decline rate when tested at the same molar dose. The same trend was observed in Figure 3.28 using Clackamas River water with PS membrane.

The effect of different doses with the Medina River water and the CA membrane can also be demonstrated using the data from the previous study, as shown in Figure 3.34. As the coagulant dose increased with this water and membrane material, the flux improved at the end of the test. These results demonstrated that clarification improved membrane performance, which was the same result observed in the current study for this source water.

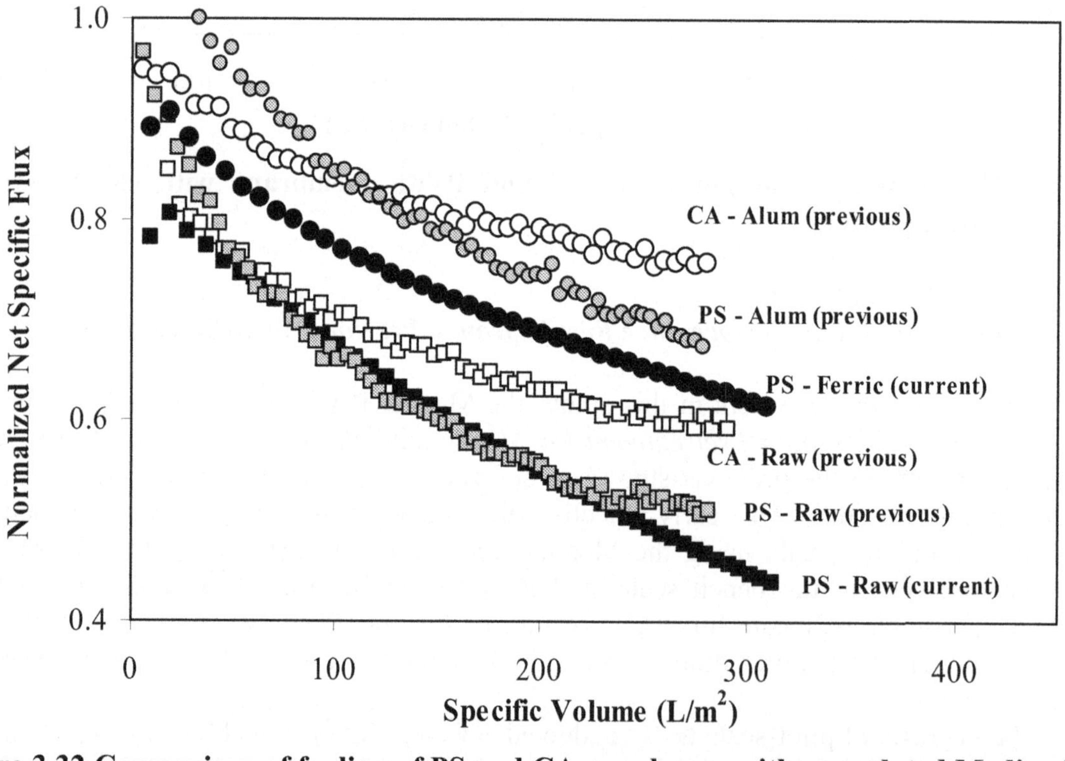

Figure 3.32 Comparison of fouling of PS and CA membrane with coagulated Medina River water

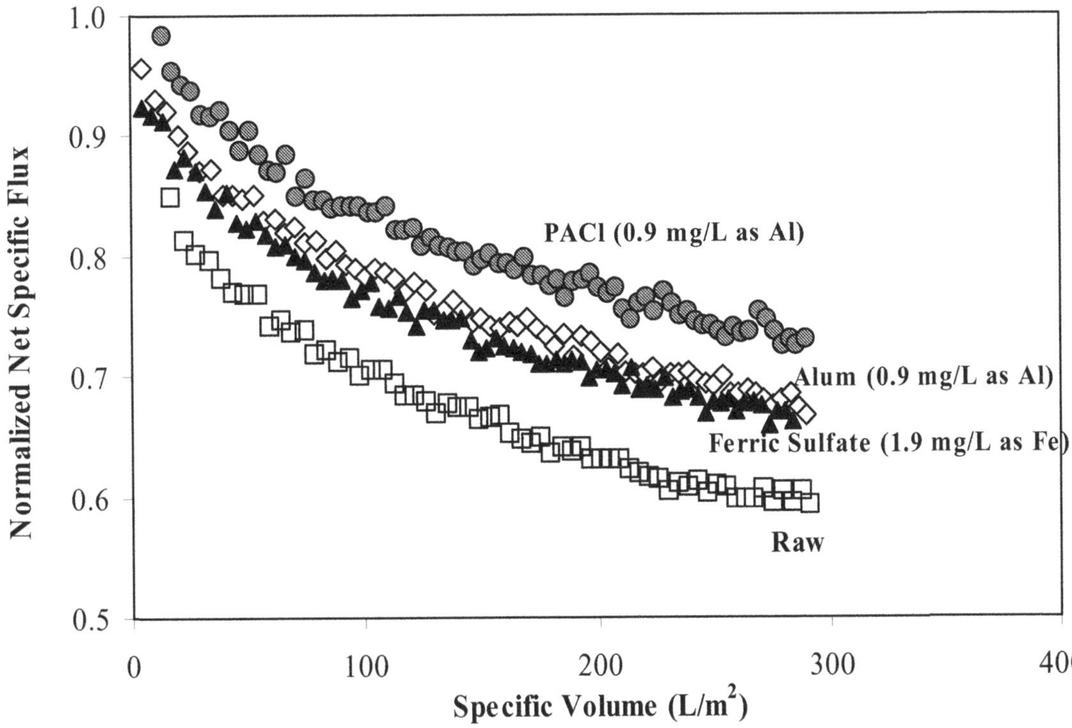

Figure 3.33 Comparison of fouling for CA membranes coagulated with PACl, alum, and ferric sulfate for Medina River water (Howe and Clark 2002a)

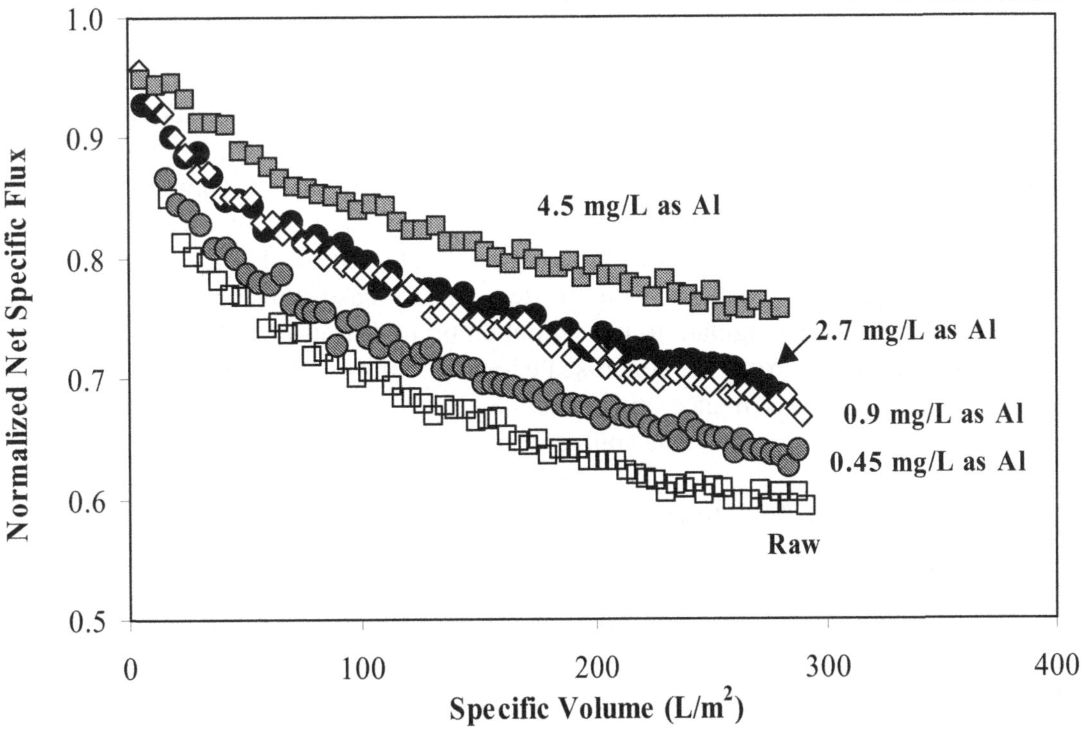

Figure 3.34 Impact of coagulant dose (alum) on CA membrane performance for Medina River water (Howe and Clark 2002a)

Summary of the Effects of Pretreatment on Membrane Flux

The effects of pretreatment on membrane performance can be summarized as follows:

1. Coagulation and clarification removed between 12 and 42 percent of the TOC in the various source waters.
2. Coagulation and clarification improved the membrane performance of the PS membranes tested with 3 different source waters. The coagulant doses were fairly high for the Medina and Seine Rivers (at 6.87 and 7.09 mg/L as Fe, respectively), which is consistent with the previous study that indicated that high coagulant doses improve membrane performance, but the Clackamas River had a low coagulant dose (0.94 mg/L as Fe) and also had improved membrane performance.
3. Coagulation and clarification did not always improve the membrane performance of the PVdF membranes with the 3 source waters tested. The Chattahoochee River water benefited from coagulation and clarification, but the Clackamas and Waikato Rivers had worse performance after pretreatment.
4. SEM imaging demonstrated a change in the foulant layer on membranes after coagulation pretreatment. Membrane surfaces were visibly cleaner or showed a even coating of flocculent material that could have offered less resistance to flow than the material seen on membranes after filtration of raw water.
5. FTIR spectrometry showed differences in membrane surface chemistry after coagulation and pretreatment, but the technique was not successful in identifying specific changes in water quality resulting from coagulation and clarification that caused changes in membrane performance.
6. PAC pretreatment on Seine River water had no impact on membrane performance at the doses tested in the bench scale tests.
7. With Seine River water, PS membrane performance improved more with alum than with ferric chloride. With Clackamas River water, however, PS membrane performance was similar with ferric sulfate and alum, and had the best performance with ACH. Specific mechanisms for these differences were not identified in this research, but it is likely that different coagulants remove different foulant species, and that the prevalence of foulant species vary between source waters. Additional research to identify specific interactions between coagulants and foulants is needed. In any case, these results suggest that the selection of the best coagulant to use with membrane filtration is site-specific.
8. At the doses tested during the bench-scale testing, chlorine pretreatment had no impact on PVdF membrane performance on Clackamas River water, but caused worse performance with PS membranes on the same source water.

Effect of Fractionation on Membrane Flux

As noted in the Materials and Methods section, for some experiments the feed waters were fractionated by prefiltering them through 1 μm, 100 kDa, or 10 kDa filters prior to performing the flux decline tests. The purpose of this fractionation was to evaluate fouling by specific components in the feed waters.

Effect of Fractionation on Fouling of PS Membranes by Raw Water

The effect of fractionation on fouling by raw waters through PS ultrafiltration membranes is shown in Figure 3.35 for the Medina River, Figure 3.36 for the Seine River, and Figure 3.37 for the Clackamas River. In each graph, the lowest net specific flux is for the unfractionated water. For the Medina and Seine Rivers, the flux after fractionation at 1 μm is nearly identical to the unfractionated water with minimal improvement on performance. These results indicate that fractionation through the 1 μm membrane had almost no effect on fouling of this membrane with these source waters, suggesting that the foulants in these waters are smaller than 1 μm in size. A greater change in flux is observed after fractionation at 1 μm in the Clackamas River (Figure 3.37), indicating that a portion of the components that foul the PS membrane in this water are larger in size. The different performance between membranes after 1 μm fractionation suggests there is a difference in the character of the foulants in the Clackamas River water compared to the other two waters.

After fractionation through the 100-kDa membranes, the net specific flux for all three source waters is higher than for unfractionated water or water fractionated at 1 μm, indicating that this fractionation was able to remove a portion of the foulants from all three source waters. Little or no additional change was observed after fractionation through the 10-kDa membranes. For all waters, regardless of the extent of fouling of the unfractionated water, the net specific flux declined to only 70 to 80 percent of the initial flux after fractionation through either the 100 kDa or 10 kDa filters. These results suggest that significant portions of the foulants are larger than 100 kDa. In summary, the foulants that affect PS membranes appear to be primarily between 1 μm and 100 kDa in size in the Medina and Seine River waters, but some are of larger size (greater than 1 μm) in the Clackamas River.

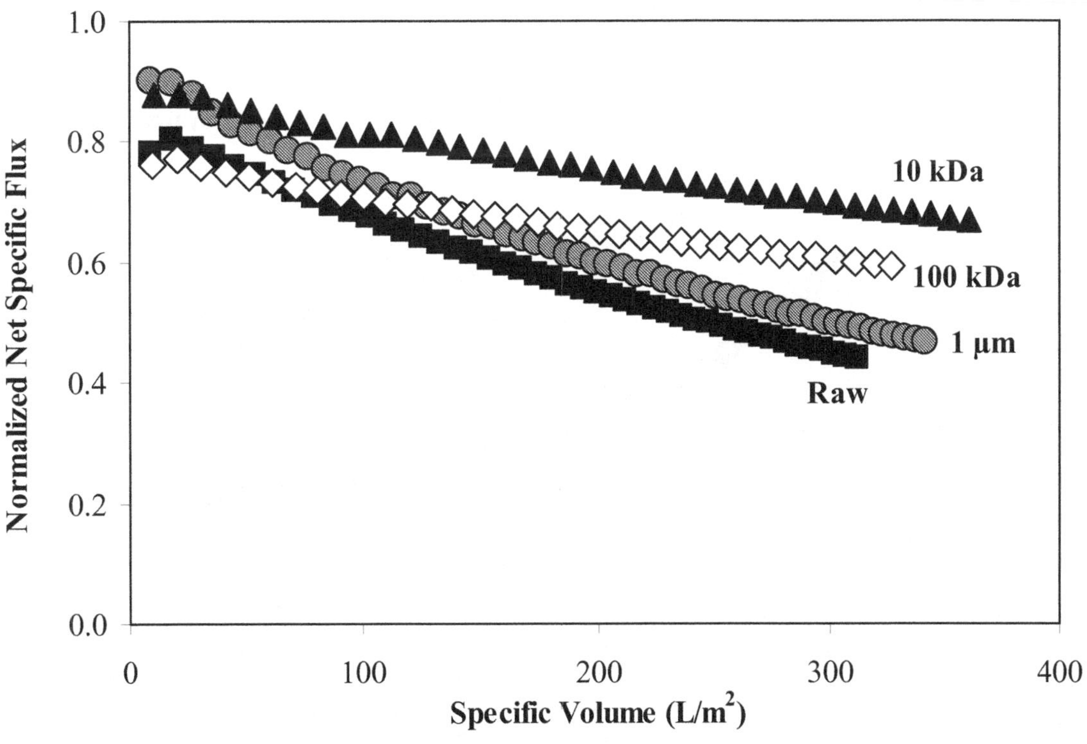

Figure 3.35 Fouling of PS membranes after fractionation of raw Medina River water

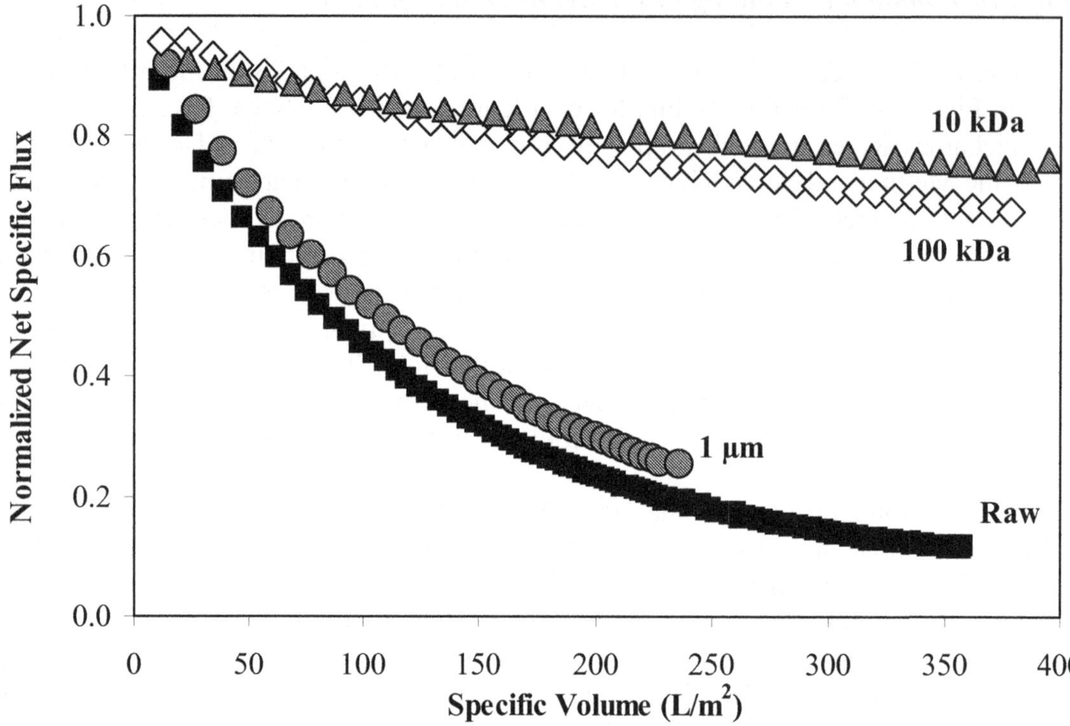

Figure 3.36 Fouling of PS membranes after fractionation of raw Seine River water

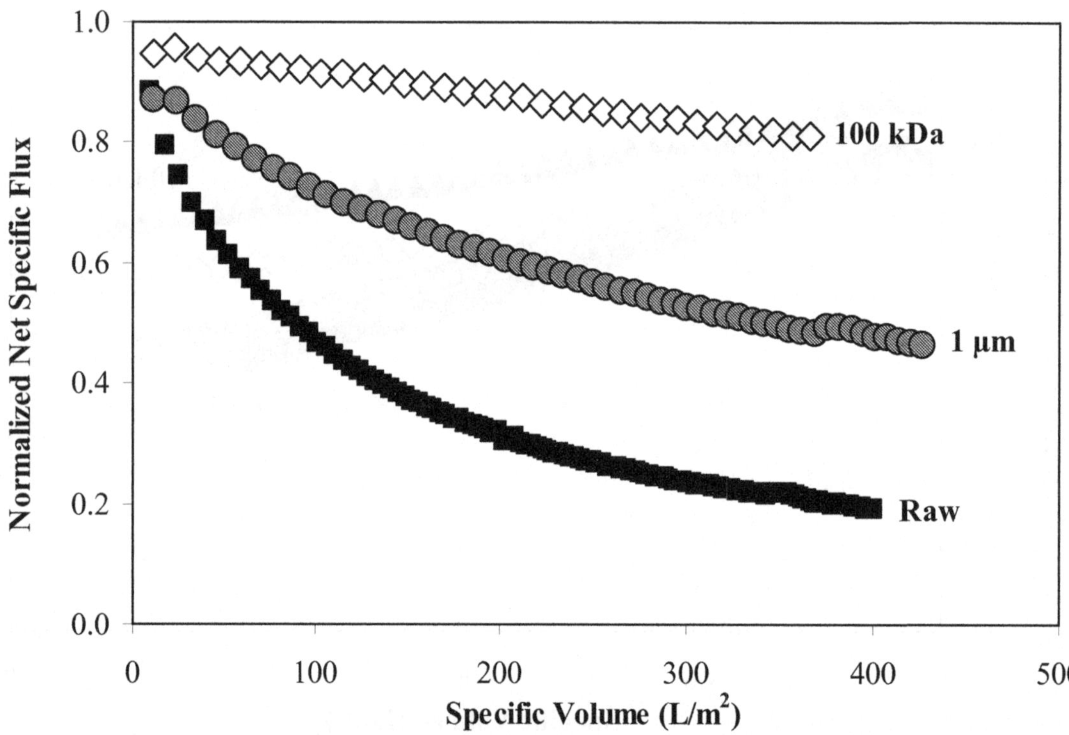

Figure 3.37 Fouling of PS membranes after fractionation of raw Clackamas River water

Effect of Fractionation on Fouling of PVdF Membranes by Raw Water

The effect of fractionation on fouling by raw waters through PVdF microfiltration membranes is shown in Figure 3.38, Figure 3.39, and Figure 3.40 for the Chattahoochee River, the Clackamas River, and the Waikato River, respectively. The effect of fractionation on the Chattahoochee River with PVdF membranes was similar to that on the Seine River with PS membranes. Fractionation though the 1 μm membrane had almost no impact on performance, fractionation through the 100 kDa dramatically reduced the net specific flux decline, and fractionation through the 10 kDa had no additional change in performance. Similar to the Seine River water with PS membrane, it appears that the predominant foulants of PVdF membranes in the Chattahoochee River are between 1 μm and 100 kDa in size.

As noted earlier, raw Clackamas River water caused very little flux decline when filtered through the PVdF membrane. As a result, fractionation through 1 μm and 100 kDa membranes had almost no change in performance, and the fractionated feed waters caused the same amount of fouling as the unfractionated water.

The effect of fractionation on the Waikato River with PVdF membranes was similar to that on the Clackamas River with PS membranes. Fractionation through the 1 μm membrane caused a significant improvement in net specific flux, and fractionation through the 100-kDa membrane generated additional improvement. These results suggest that a portion of the foulants in this water that affect PVdF membranes are larger than 1 μm, and a portion are between 1 μm and 100 kDa in size.

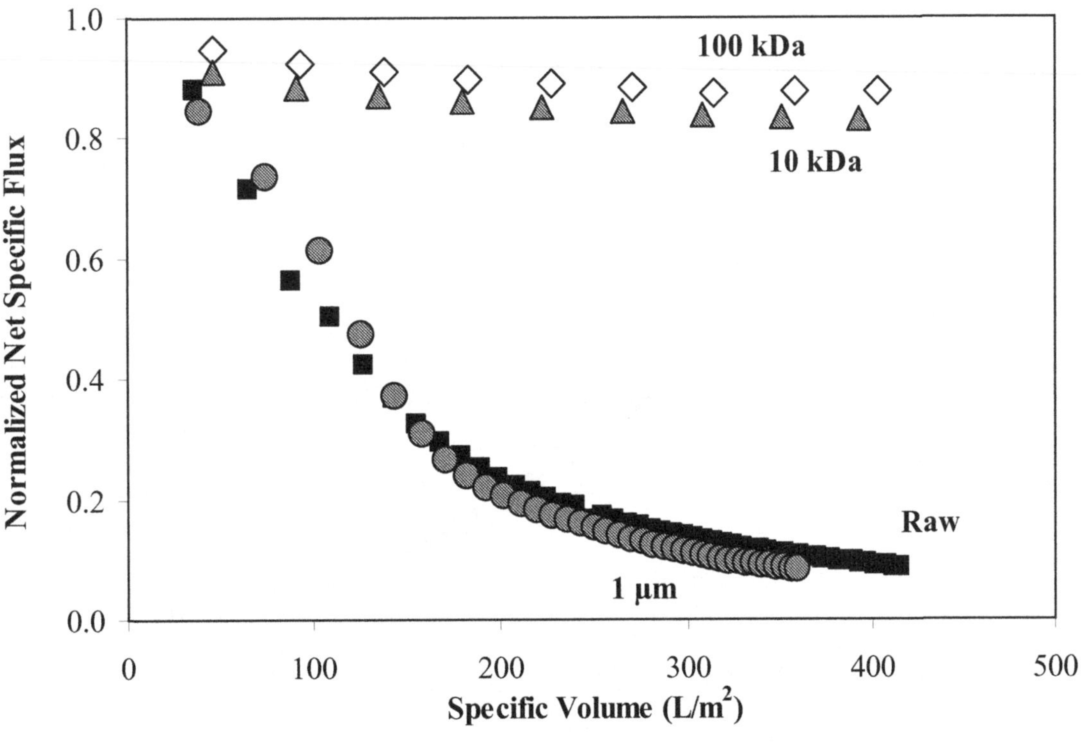

Figure 3.38 Fouling of PVdF membranes after fractionation of raw Chattahoochee River water

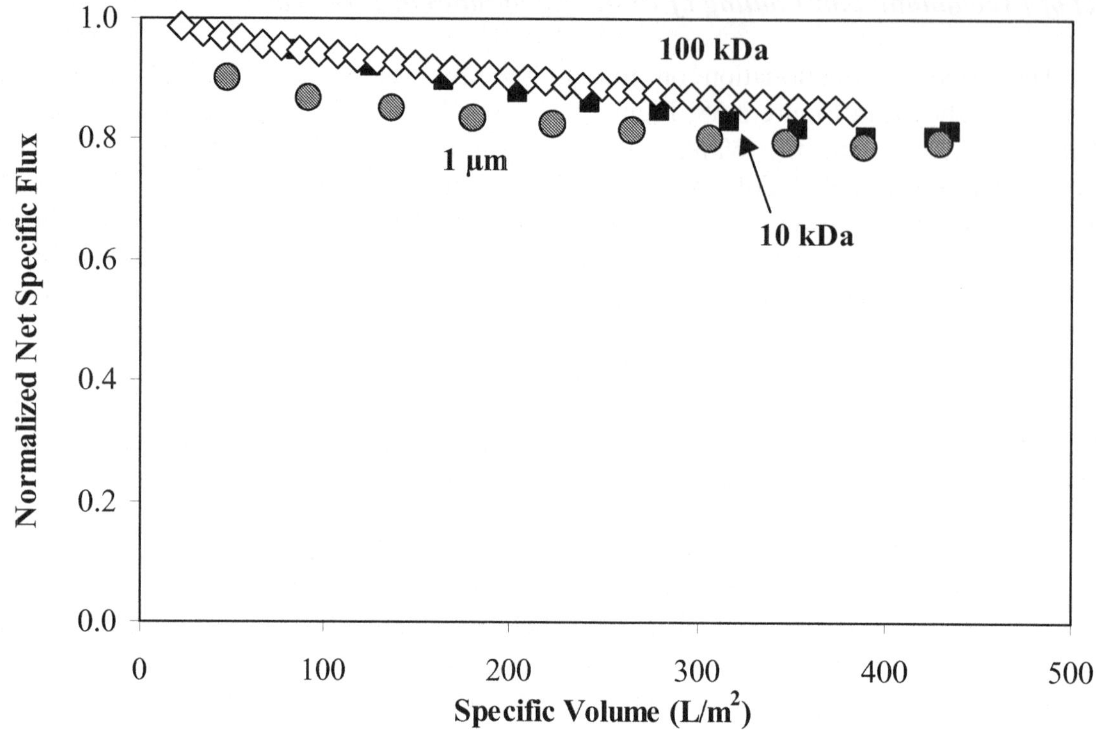

Figure 3.39 Fouling of PVdF membranes after fractionation of raw Clackamas River water

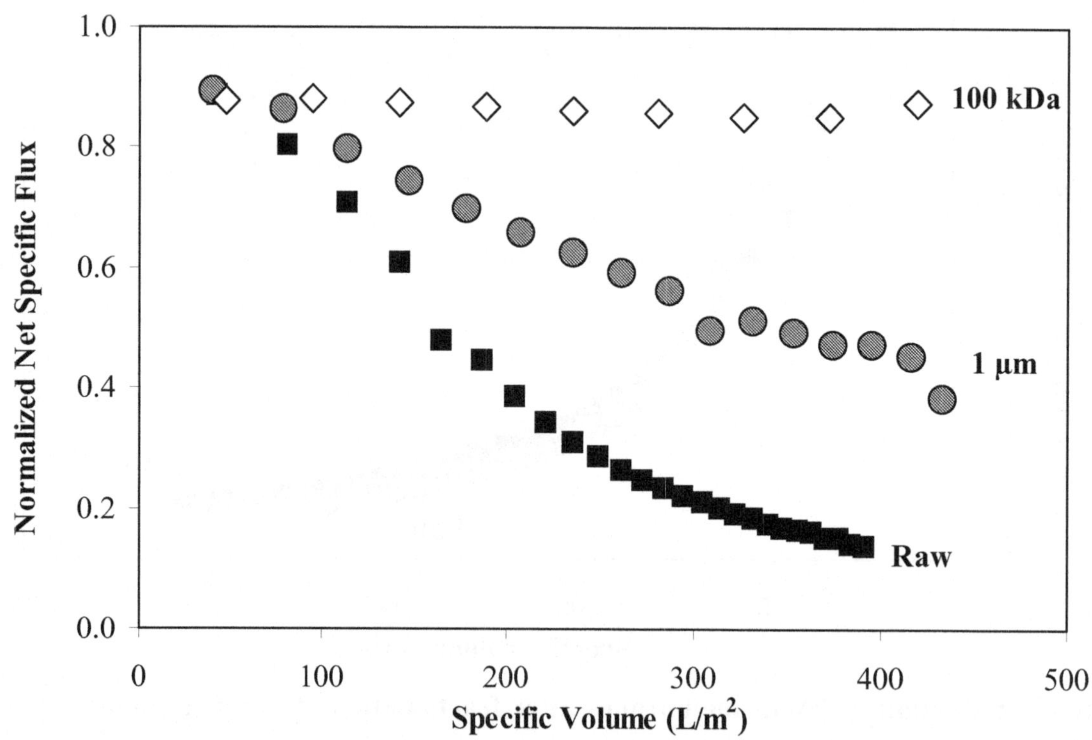

Figure 3.40 Fouling of PVdF membranes after fractionation of raw Waikato River water

SEM Characterization of Membranes Fouled with Fractionated Raw Water

SEM images of PVdF membranes fouled with Waikato River water after fractionation through 1 μm and 100 kDa membranes are shown in Figure 3.41 and Figure 3.42. These images can be compared to the image of a PVdF membrane fouled with raw Waikato River water (Figure 3.11). In Figure 3.41, some deposition can be observed on the surface of the membrane, but it is less than the deposition observed in Figure 3.11. The reduction in deposition in these two images is consistent with the reduction in fouling observed in Figure 3.40. Almost no deposition is observed in Figure 3.42, and the image looks similar to the image of an unused PVdF membrane (Figure 3.10). These images suggest that the smaller material responsible for fouling the PVdF membranes after 1 μm fraction were successfully removed by the 100 kDa membranes.

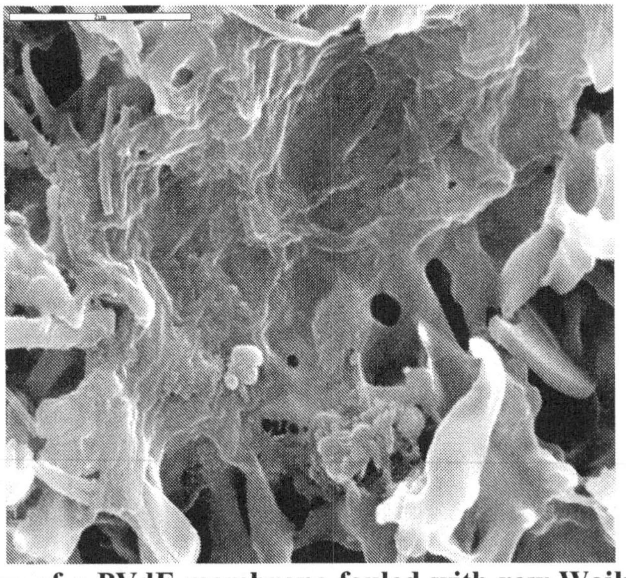

Figure 3.41 SEM image of a PVdF membrane fouled with raw Waikato River water after fractionation through a 1 μm membrane

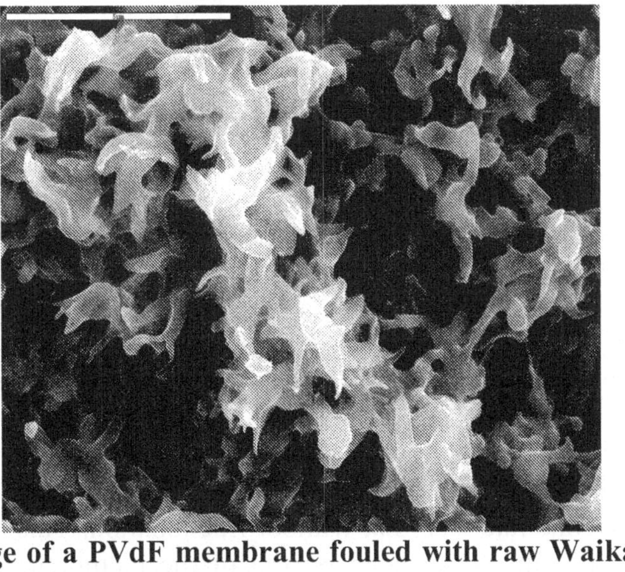

Figure 3.42 SEM image of a PVdF membrane fouled with raw Waikato River water after fractionation through a 100 kDa membrane

FTIR Characterization of Membranes Fouled with Fractionated Raw Water

FTIR spectra of raw Waikato River water with and without fractionation through a 1 μm membrane are shown in Figure 3.43. The solid black line in the graph shows the spectrum from the membrane fouled with raw water and the gray line is the spectrum of the membrane fouled with fractionated water. The raw water spectrum shows a large peak near 1000 cm^{-1}, which was noted earlier to be due to C-O or Si-O bonds, indicating the possible presence of silica on the membrane surface. After fractionation through the 1 μm membrane, the peak at 1000 cm^{-1} is substantially diminished. This suggests that bigger particles that were removed from the sample water by fractionation may have had absorbance in this region, and could have a silica-based component to it. The remaining significant peaks after 1 μm fractionation are at 1400, 1180, and 875 cm^{-1}. These peaks could be caused by a variety of organic compounds, including aliphatic and aromatic nitrogenous compounds, sulfones and sulfonates, and mononuclear aromatic compounds.

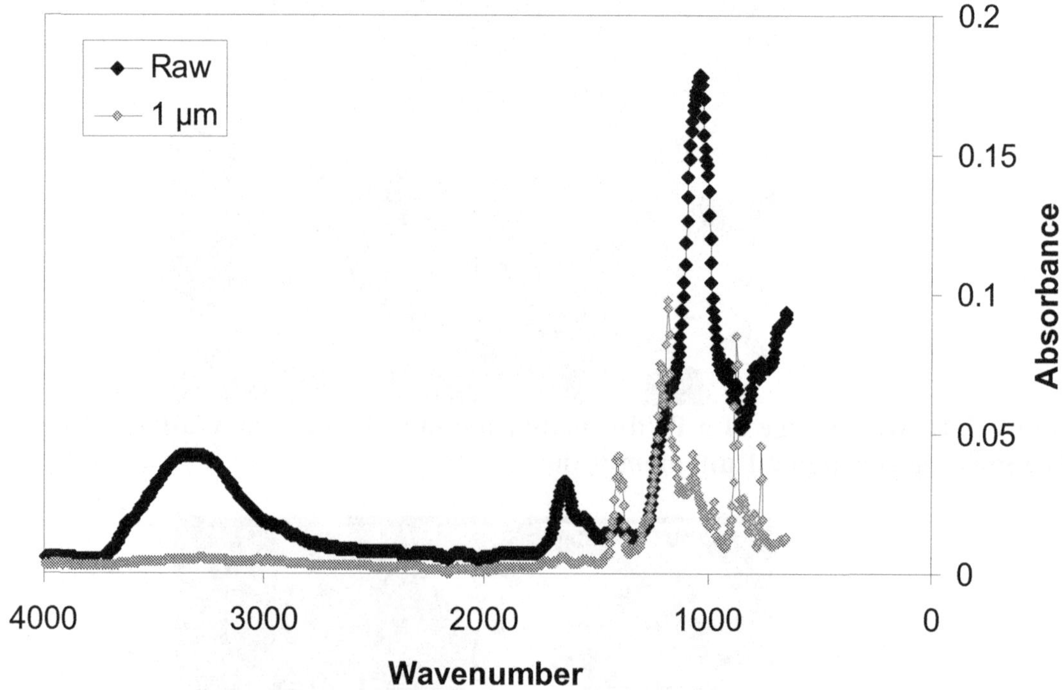

Figure 3.43 FTIR spectra of PVdF membranes fouled with raw Waikato River water with and without 1 μm fractionation

Effect of Fractionation on Fouling of PS Membranes by Clarified Water

The effect of fractionation on fouling of PS ultrafiltration membranes by clarified Medina River water is shown in Figure 3.44. The net specific flux after fractionation through the 1 μm membrane has faster decline rate than that without fractionation. This observation suggests that larger floc particles can form a cake layer on top of the membrane, which can then remove smaller foulants. When the cake layer of larger particles is eliminated by prefiltering the feed water through a 1μm fractionation membrane, smaller particles encountered the membrane and fouled the membrane. After fractionation at 100 kDa, the net specific flux of clarified Medina River water was similar to that after the same fractionation of raw water. These results suggest that coagulation and clarification removed foulants between 1 μm and 100 kDa in size (improving the unfractionated flux) but had minor impact on the portion of the foulants less than 100 kDa in size.

A similar trend is shown for the fouling of clarified Seine River water in Figure 3.45. The 1 μm fractionated clarified water shows a drop in flux compared to the clarified not-fractionated water. The 100-kDa fraction shows a greater increase in flux, pointing out that the clarification of Seine River raw water removed less of the foulants above 100 kDa than the clarification for Medina River water.

Fractionation of Clackamas River water with 1-μm membranes after coagulation with ferric and alum, as shown in Figure 3.46, did not produce the effect that was seen in the Medina and Seine Rivers. This observation is additional evidence that the foulant species in the Clackamas River water had different characteristics than the foulant in the other waters. The cake layer produced by coagulation and clarification of Clackamas River water appeared to contribute to fouling of the membrane, so that removal of a portion of the cake layer by 1 μm fractionation slightly reduced fouling, whereas an increase in fouling was observed with the other two source waters.

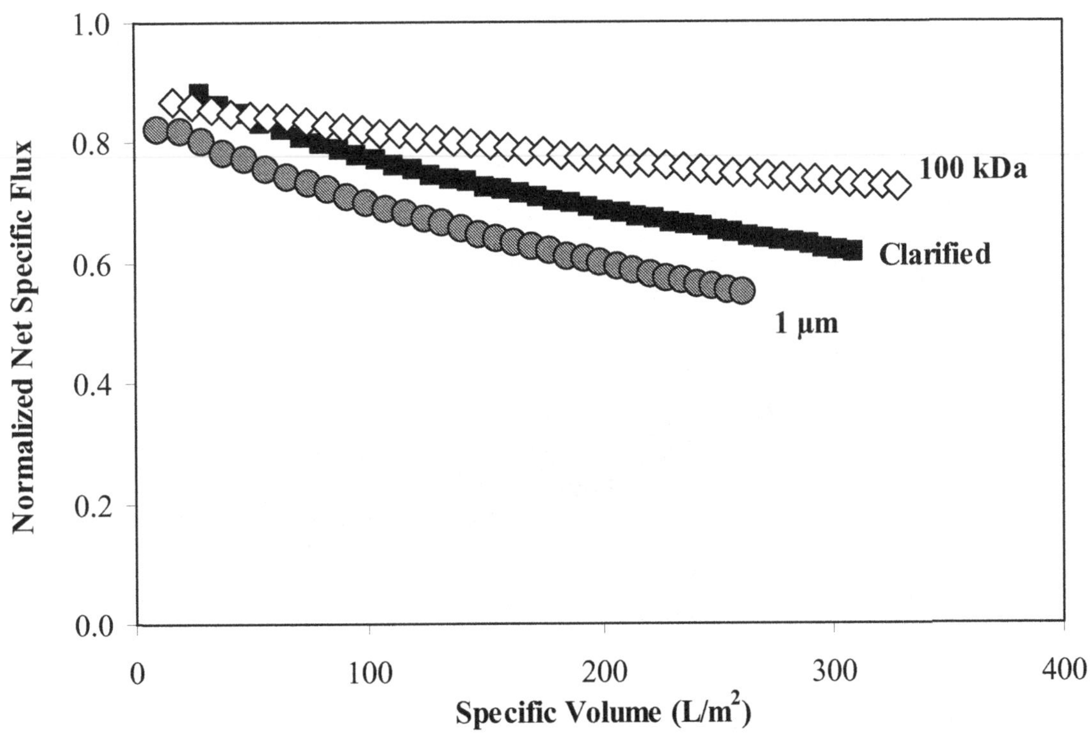

Figure 3.44 Fouling of PS membranes after fractionation of ferric-clarified Medina River water

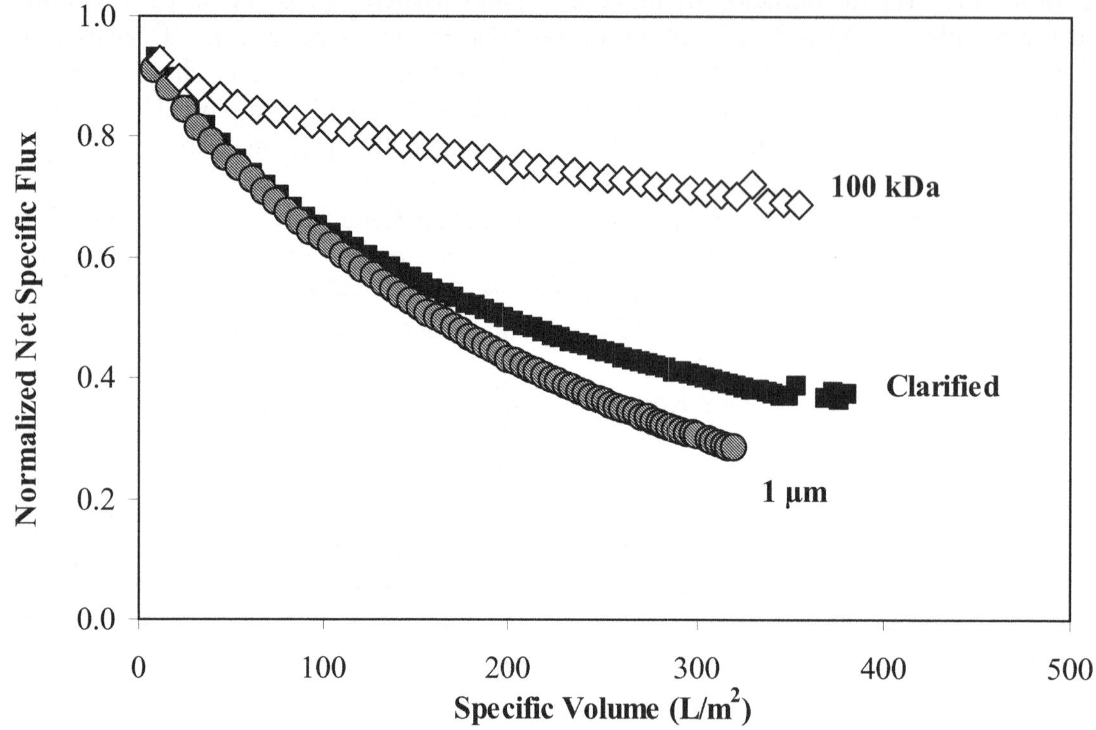

Figure 3.45 Fouling of PS membranes after fractionation of ferric-clarified Seine River water

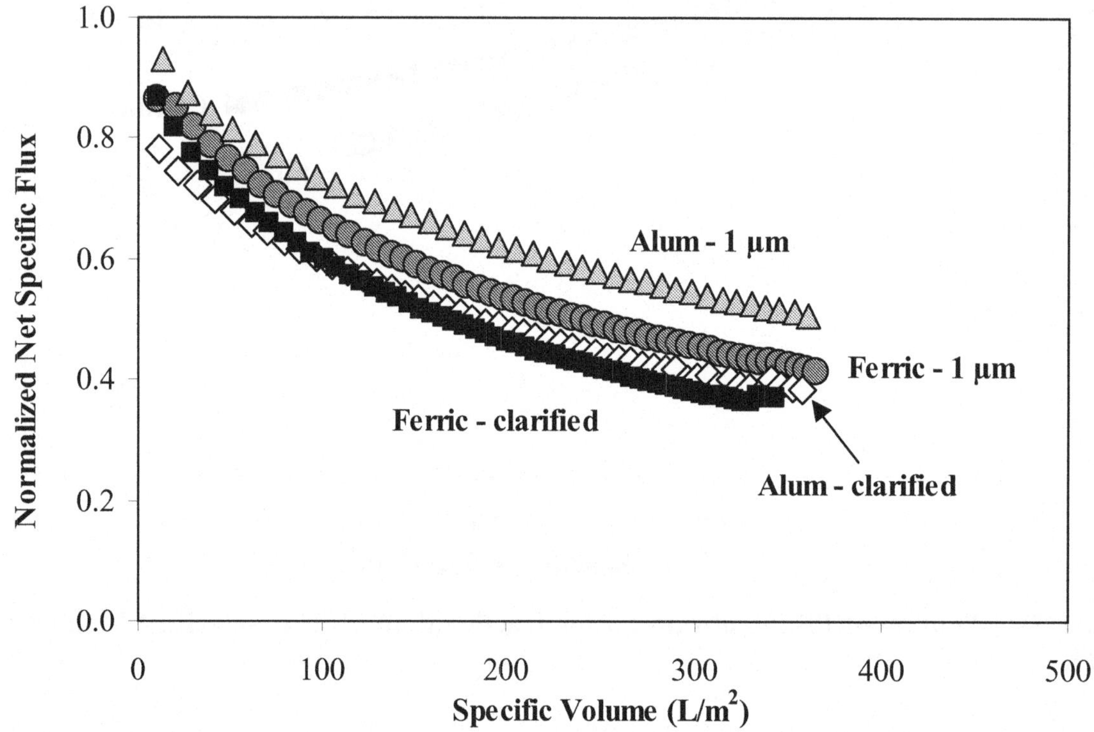

Figure 3.46 Fouling of PS membranes after fractionation of clarified Clackamas River water

Effect of Fractionation on Fouling of PVdF Membranes by Clarified Water

The three curves in Figure 3.47 shows the net specific flux of clarified Chattahoochee River water after prefiltration through 1 µm and 100 kDa membranes, using PVdF microfiltration membranes and alum as the coagulant. The clarified Chattahoochee River water had a substantial improvement in flux after prefiltration through the 1µm membrane, but no additional improvement after prefiltration through the 100 kDa membrane. This suggests that the material in clarified Chattahoochee River water that fouled the PVdF membrane was larger than 1 µm. Smaller material appears to have passed through without fouling the membrane.

Fractionation of Clackamas River water with alum as the coagulant and filtration through the PVdF membranes is shown in Figure 3.48. The figure consists of three curves with a pattern similar to Figure 3.47. The alum-clarified water fouled the membrane extensively (recall that the clarification increased fouling compared to raw Clackamas River water using PVdF microfiltration membranes in Figure 3.20). When this clarified water was fractionated with 1 µm glass filter, the flux shows substantial improvement suggesting that the larger particles formed after coagulation are responsible for fouling of PVdF membrane. After fractionation with the 100-kDa membranes, this water does not show any fouling on the PVdF membrane.

Fractionation of Clackamas River water with ACH as the coagulant and filtration through the PVdF membranes is shown in Figure 3.49. Compared to alum, ACH was a better coagulant for the Clackamas River water. For ACH-clarified water, when the sample was fractionated with 1 µm glass filter, the foulant seems to be removed as the flux increases and reaches a flat curve with little or no fouling. This observation indicates that clarification with different coagulant chemical result in different composition of foulants hence membrane performance.

Fractionation of Waikato River water with alum as the coagulant and filtration through the PVdF membranes is shown in Figure 3.50. The flux graphs show that 1 µm fractionation removed a substantial amount of the foulants. As noted earlier, coagulation and clarification made fouling worse in the Waikato River water (Figure 3.20, Waikato River raw and clarified flux). Similar to the Clackamas River water clarified with ACH as coagulant, a significant portion of material that fouled PVdF membranes after coagulation and clarification appeared to be larger than 1 µm in size, since fractionation of this clarified water through the 1µm membranes reduced fouling and increased flux.

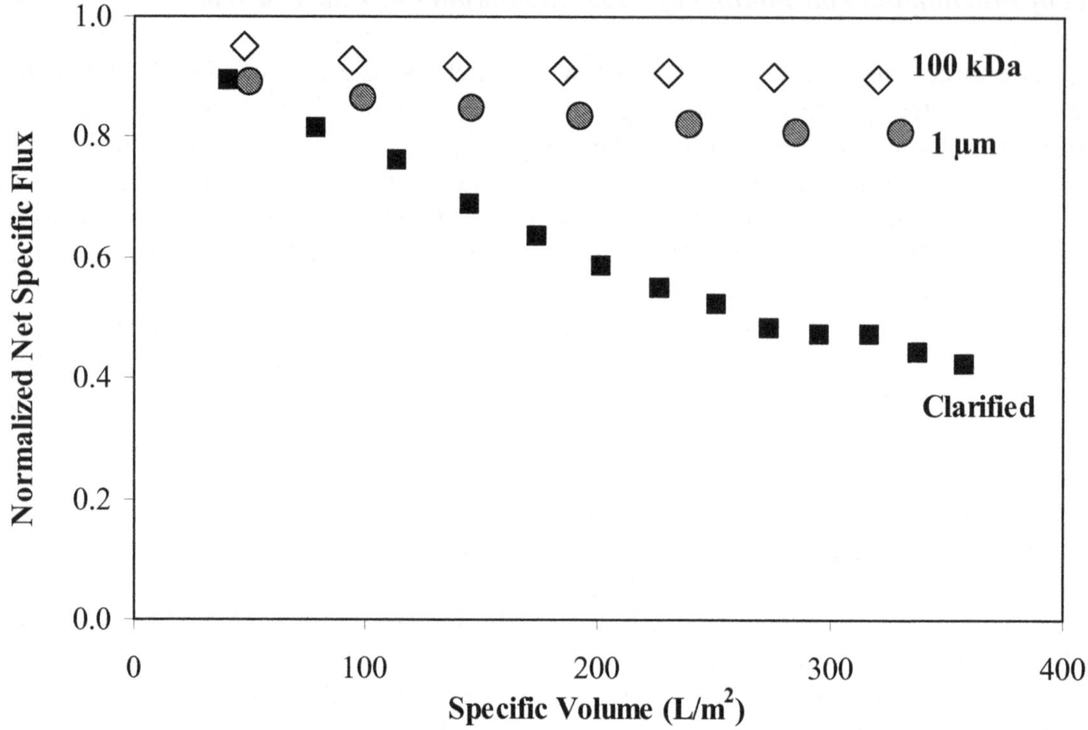

Figure 3.47 Fouling of PVdF membranes after fractionation of alum-clarified Chattahoochee River water

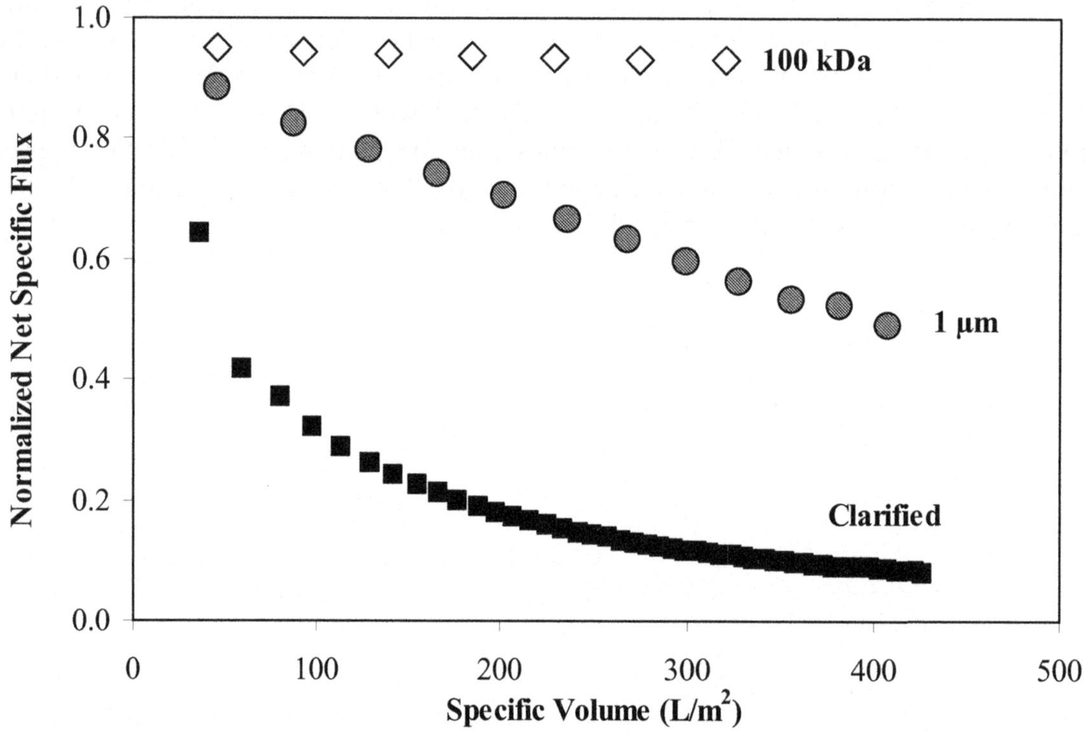

Figure 3.48 Fouling of PVdF membranes after fractionation of alum-clarified Clackamas River water

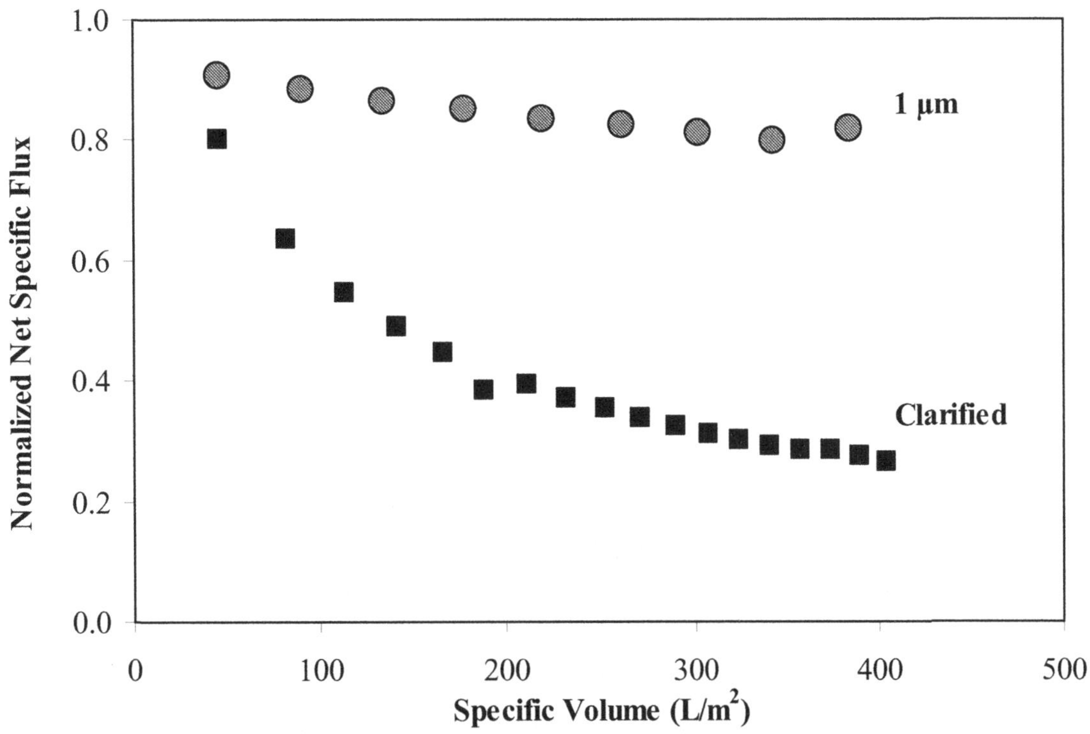

Figure 3.49 Fouling of PVdF membranes after fractionation of ACH-clarified Clackamas River water

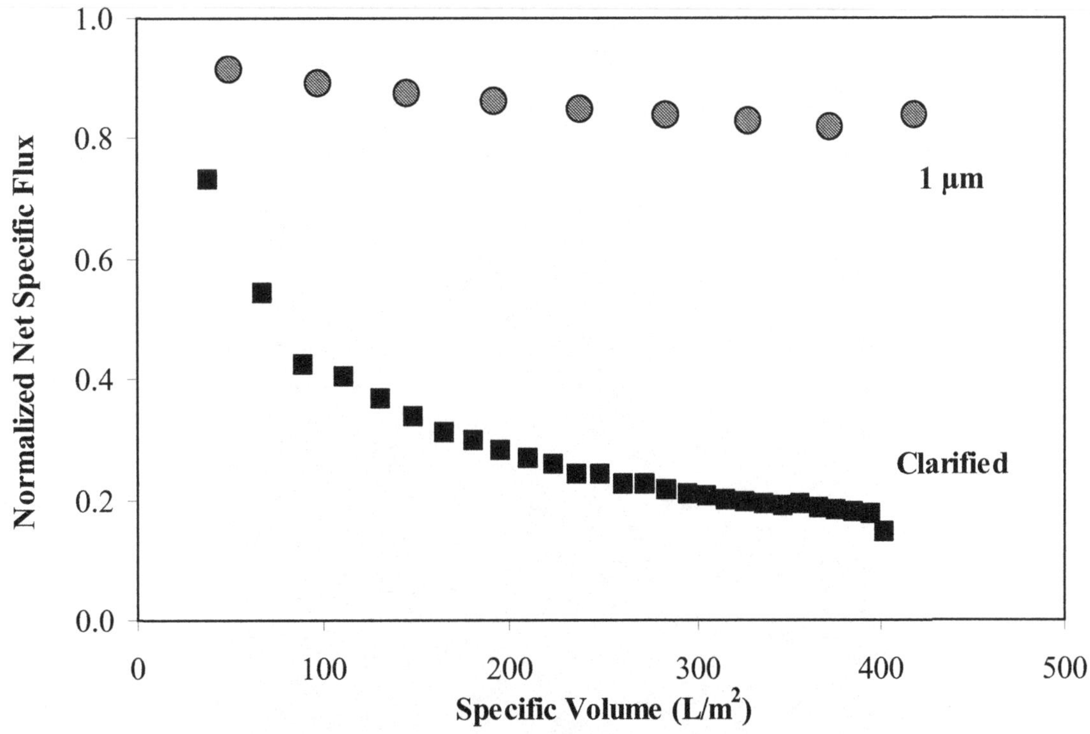

Figure 3.50 Fouling of PVdF membranes after fractionation of alum-clarified Waikato River water

SEM Characterization of PVdF Membranes Fouled with Fractionated Clarified Water

An SEM image of a PVdF membrane fouled with clarified Waikato River water with alum as coagulation is shown in Figure 3.51. This image shows heavy deposition on the surface of the membrane, similar to what was observed with raw Waikato River water in Figure 3.11. As noted earlier of specific flux decline (Figure 3.20), the fouling of raw Waikato River water was as significant as it was of clarified water. A comparison of the SEM images in Figure 3.11 and Figure 3.51 suggests that the surface cake layer is responsible for the fouling in this source water. After fractionation through a 1-μm membrane, the SEM image of the surface of the PVdF membrane is substantially different, as shown in Figure 3.52. The image does not show deposition on the surface of the membrane, and the image is similr to the unused PVdF membrane (Figure 3.10). This observation is consistent with the flux shown in Figure 3.50, where no fouling is observed after 1 μm fractionation of alum-clarified water.

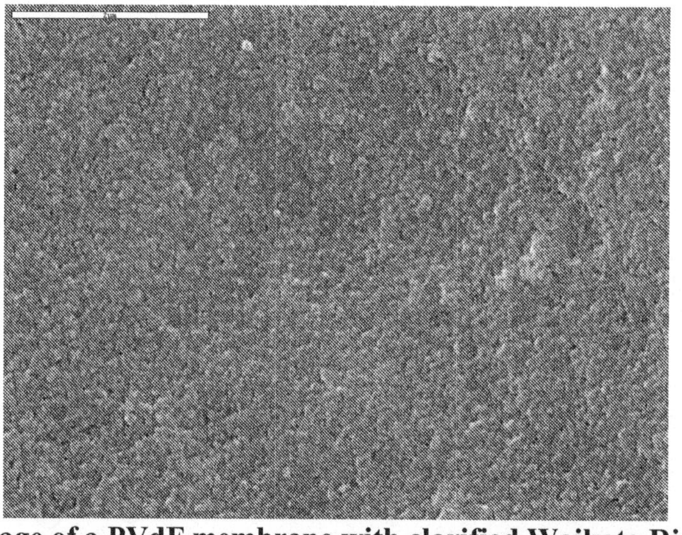

Figure 3.51 SEM image of a PVdF membrane with clarified Waikato River water

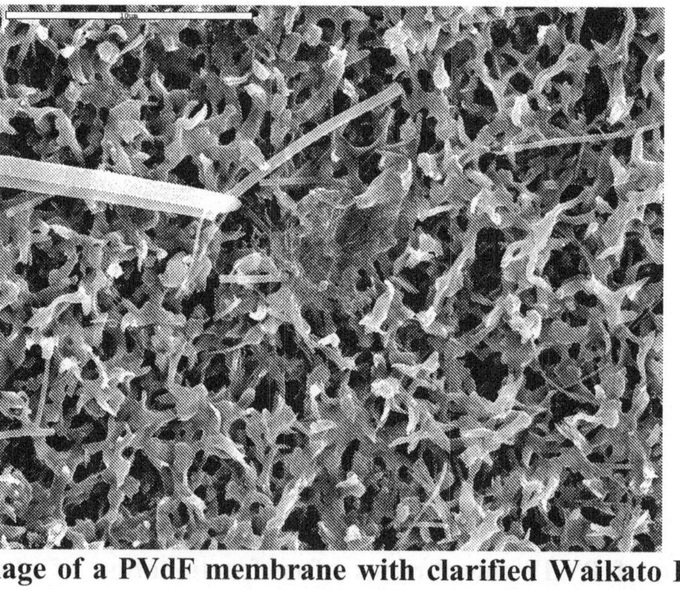

Figure 3.52 SEM image of a PVdF membrane with clarified Waikato River water after 1 μm fractionation

Summary of the Effect of Size Distribution on Membrane Fouling

The effects of the fractionation experiments on membrane flux are summarized in Table 3.11. The primary conclusions from this portion of the research can be summarized as follows:

1. Size fractionation provided useful insight into the nature of foulants from various source waters.
2. Prefiltration of raw water through 1 μm membranes made almost no change in membrane performance for the Chattahoochee and Clackamas River waters using PVdF membranes and the Medina and Seine waters using PS membranes. Membrane performance (measured as normalized net specific flux) improved after Clackamas (using PS membranes) and Waikato River water (using PVDF membranes) was prefiltered through the 1.0 μm membranes. This suggests that the major foulants were below 1 μm in size in the Medina, Seine, and Chattahoochee Rivers, but that some foulants were larger than 1 μm in the Clackamas and Waikato Rivers. The difference in performance of the Clackamas River water with the two types of membranes may be due to differences in the size of the foulants.
3. Prefiltration of raw water through 100 kDa membranes improved the flux through all membranes with all source waters (except the Clackamas River with PVdF membranes, where almost no fouling was observed with raw water). In fact, normalized flux was 60 to 85 percent of the clean membrane flux after prefiltration at this level for all source waters. This suggests that a major fraction of membrane foulants is larger than 100 kDa in all of the waters tested, although some foulants are clearly smaller than the 10 kDa size. Additional prefiltration at 10 kDa caused little or no additional improvement in flux, suggesting that there is little or no material between 10 and 100 kDa in size that contributes to membrane fouling.
4. FTIR spectrometry noted a substantial reduction of a peak near 1000 cm^{-1} after fractionation, which could indicate removal of a silica-based compound during prefiltration.
5. For clarified water and PS ultrafiltration membranes, prefiltration through the 1 μm membranes caused worse performance than unfractionated water for the clarified Medina and Seine River waters. These results suggest that the floc larger than 1 μm in the coagulated water created a cake layer on the membrane, which provided a removal mechanism for smaller foulants and prevented them from reaching the membrane. Prefiltration removed the floc larger than 1 μm, restricted the formation of the cake layer, allowed the smaller material to reach the membrane, and caused additional fouling. This effect was not seen with Clackamas River water, however, where a slight improvement was seen following clarification. The cake layer formed by the Clackamas River water may have been less effective at capturing smaller particles.
6. After prefiltration through the 1 μm membranes, PVdF membranes had substantially improved performance after filtration of clarified water. This result was true for several source waters (Chattahoochee, Clackamas, and Waikato) and different coagulants (alum and ACH). The difference in performance between the PS and PVdF membranes suggests different mechanisms of fouling for these UF and MF membranes. The PVdF, which was a MF membrane, appears to be fouled primarily by particles larger that 1 μm

after the feed water is coagulated and clarified. These larger flocculated particles may clog the pores of a MF membrane, but form a cake layer on the UF membranes.

7. After prefiltration through the 100 kDa membranes, all membranes experienced improved performance using coagulated/clarified feed waters with all source waters. The membranes were able to maintain 70 to 90 percent of the original flux, which was similar to their performance during filtration of raw water that was prefiltered through 100 kDa membranes. These results suggest that coagulation and clarification may have little or no impact on foulants smaller than 100 kDa in size. Coagulation/sedimentation research supports the hypothesis that coagulation preferentially removes larger material.
8. SEM imaging supported the flux results from fractionation of raw and clarified water. Fouled membranes appeared relatively similar to unfractionated membranes after prefiltration through the 1 μm membrane, but appeared relatively clean after fractionation through the 100-kDa membranes, which is consistent with the flux results.

Table 3.11
Summary of the effect of size distribution on membrane fouling

	PVDF Membranes			**PS Membranes**		
	Chattahoochee	**Clackamas**	**Waikato**	**Medina**	**Seine**	**Clackamas**
Raw water	Flux declined to 10% of clean membrane flux (J/Jo = 10%)	Flux declined to 80% of clean membrane flux (J/Jo = 80%)	Flux declined to 10% of clean membrane flux (J/Jo = 10%)	Flux declined to 43% of clean membrane flux (J/Jo = 43%)	Flux declined to 10% of clean membrane flux (J/Jo = 10%)	Flux declined to 18% of clean membrane flux (J/Jo = 18%)
1 μm prefiltration	No change in flux from raw	No change in flux from raw	Increased J/Jo to 40%	Slight increase in J/Jo	Slight increase in J/Jo	Increased J/Jo to 45%
100 kDa prefiltration	Increased J/Jo to 80%	No change in flux from raw	Increased J/Jo to 85%	Increased J/Jo to 60%	Increased J/Jo to 70%	Increased J/Jo to 82%
10 kDa prefiltration	No change in flux from 100 kDa	N/A	N/A	Slight increase in J/Jo from 100 kDa	Slight increase in J/Jo from 100 kDa	N/A
Clarification	J/Jo increased compared to raw (J/Jo = 40%)	J/Jo decreased compared to raw (J/Jo = 10% for alum, J/Jo = 25% for ACH)	J/Jo increased compared to raw (J/Jo = 20%)	J/Jo increased compared to raw (J/Jo = 60%)	J/Jo increased compared to raw (J/Jo = 35%)	J/Jo increased compared to raw (J/Jo = 35% for both alum and ferric)
1 μm prefiltration	Increased J/Jo to 80%	Increased J/Jo to 50% for alum, 80% for ACH	Increased J/Jo to 80%	J/Jo lower than with ferric-clarified water	J/Jo lower than with ferric-clarified water	Slight increase in J/Jo with both alum and ferric
100 kDa prefiltration	Increased J/Jo to 90 %	Increased J/Jo to 90% for alum, N/A for ACH	N/A	Increased J/Jo to 70%	Increased J/Jo to 70%	N/A

Comparison of Flux of Flat Sheet and Hollow Fiber Membranes

Bench scale tests were conducted using both flat sheet and hollow fiber membranes. Initially, only a small number of hollow fiber membrane experiments were planned, with the purpose of confirming that the flux decline characteristics of flat sheet membranes used at bench scale were similar to hollow fiber membranes used at larger scale. The initial experimentation, however, revealed substantial differences between the performance of flat sheet and hollow fiber membranes, and a larger number of experiments were conducted.

Flat Sheet and Hollow Fiber Comparison with PVdF Microfiltration Membranes

The difference between flat sheet and hollow fiber configurations was investigated with the PVdF microfiltration membranes. Raw Chattahoochee River and Clackamas River water were filtered through flat sheet and hollow fiber membranes as shown in Figure 3.53 and Figure 3.54, respectively. For both source waters, the flat sheet membrane was a commercial product, the PVdF membrane from Millipore, which is reported to be a hydrophilized product. The hollow fiber membranes (PVdF1) used with the Chattahoochee River water were provided by US Filter, and was considered the more hydrophobic of the PVdF hollow fiber products that were provided. In Figure 3.53, the hollow fiber shows considerably less fouling than the flat sheet membrane. Because the two products were provided by different manufacturers, the difference in fouling could be due to differences in surface properties of the membranes, even though they were nominally the same material. The difference in fouling could be caused by structural difference due to difference in fabrication methods, differences in clean water permeability, or differences in the mechanism of fouling.

The hollow fiber membrane (PVdF2) shows more fouling than the flat sheet membrane with the Clackamas River water, as shown in Figure 3.54, but the difference is less dramatic than in the previous figure. Although the hollow fiber membrane used in these tests was a newer product that was reported to be more hydrophilic, it cannot be concluded that a greater similarity in membrane materials was responsible for the better agreement between the flat sheet and hollow fiber products. A primary cause of the smaller difference in fouling with the Clackamas River was the relatively low amount of fouling observed with PVdF microfiltration membranes with this source water. Of the numerous comparisons between hollow fiber and flat sheet membranes, this was the only instance where the fouling of the hollow fiber membrane was greater than that of the flat sheet membrane. In all other cases, the fouling of the flat sheet membrane was greater than that of the hollow fiber membrane.

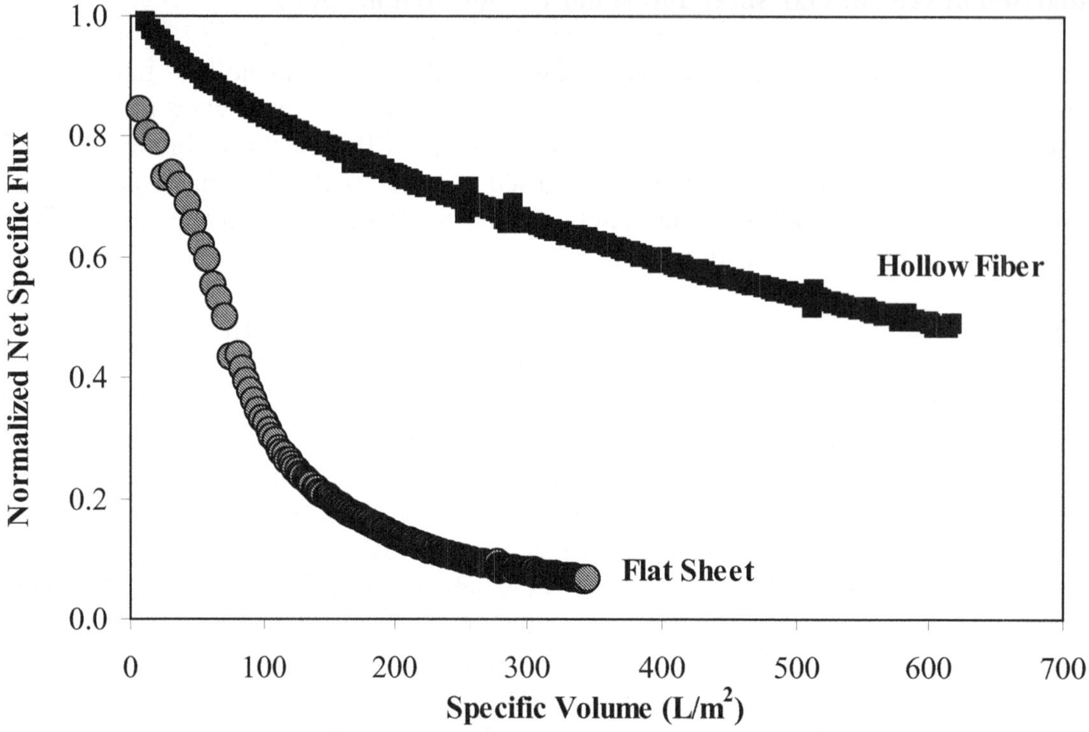

Figure 3.53 Fouling of PVdF hollow fiber and flat sheet membranes during filtration of raw Chattahoochee River water

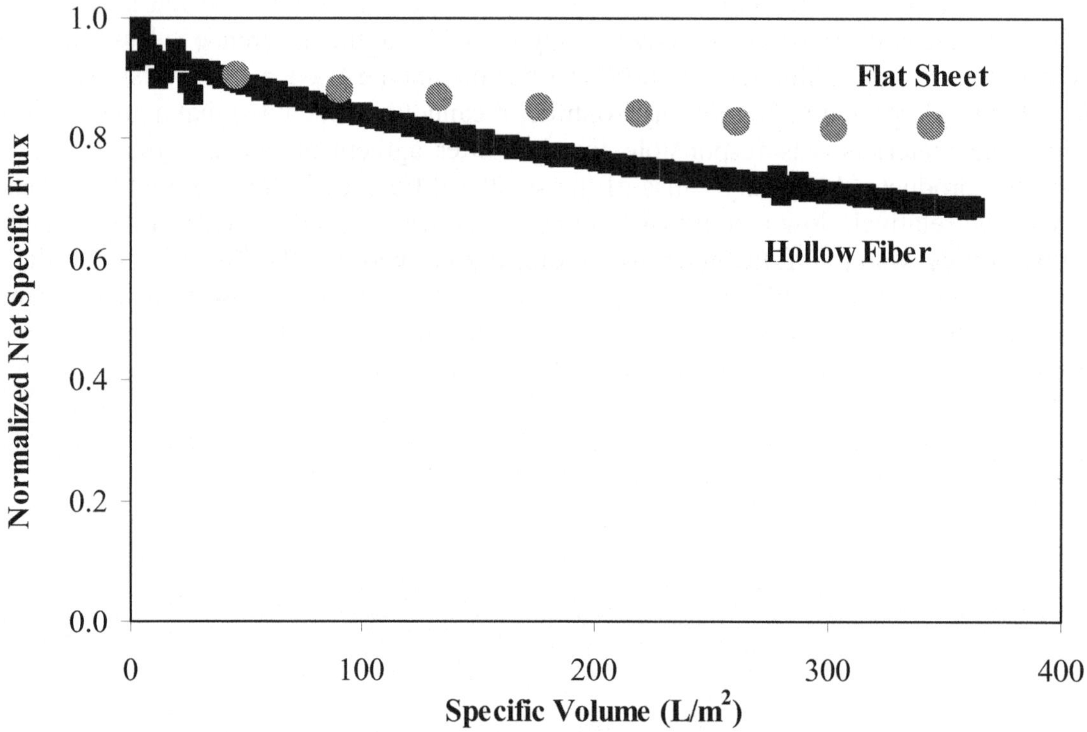

Figure 3.54 Fouling of PVdF hollow fiber and flat sheet membranes during filtration of raw Clackamas River water

Flat Sheet and Hollow Fiber Comparison with PS Ultrafiltration Membranes

A comparison of the fouling behavior of hollow fiber and flat sheet membrane during filtration of raw Clackamas river water is shown in Figure 3.55. This figure shows that hollow fiber membranes have less fouling than flat sheet membranes, even when both are nominally the same material and filtering the same source water. The difference of clean water permeability between flat sheet and hollow fiber PS membranes is much lower than that of PVdF membranes presented in previous section; therefore, the different membrane performance between flat sheet and hollow fiber is unlikely caused by clean water permeability difference.

For these experiments, the flat sheet and hollow fiber membranes were both provided by Aquasource, and were fabricated of the same material. The hollow fiber membranes were the commercially available product and were fabricated on an industrial scale, while the flat sheet membranes were fabricated by hand by research staff. The differences in fouling could have been influenced by structural differences resulting from the difference in fabrication methods, or from different mechanisms in fouling due to the difference in configuration. A similar result is observed with flat sheet and hollow fiber membranes being used to filter raw Waikato River water in Figure 3.56. The PS flat sheet membranes show a higher degree of fouling than the hollow fiber membranes.

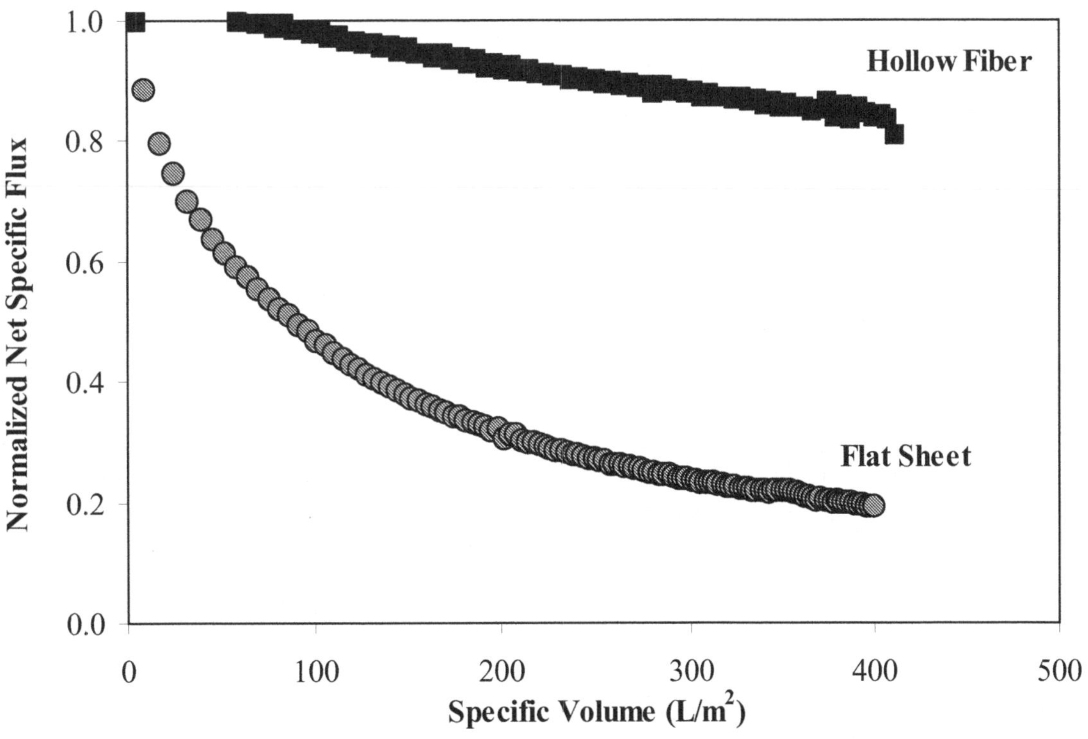

Figure 3.55 Flux of raw Clackamas River water through flat sheet and hollow fiber PS ultrafiltration membranes

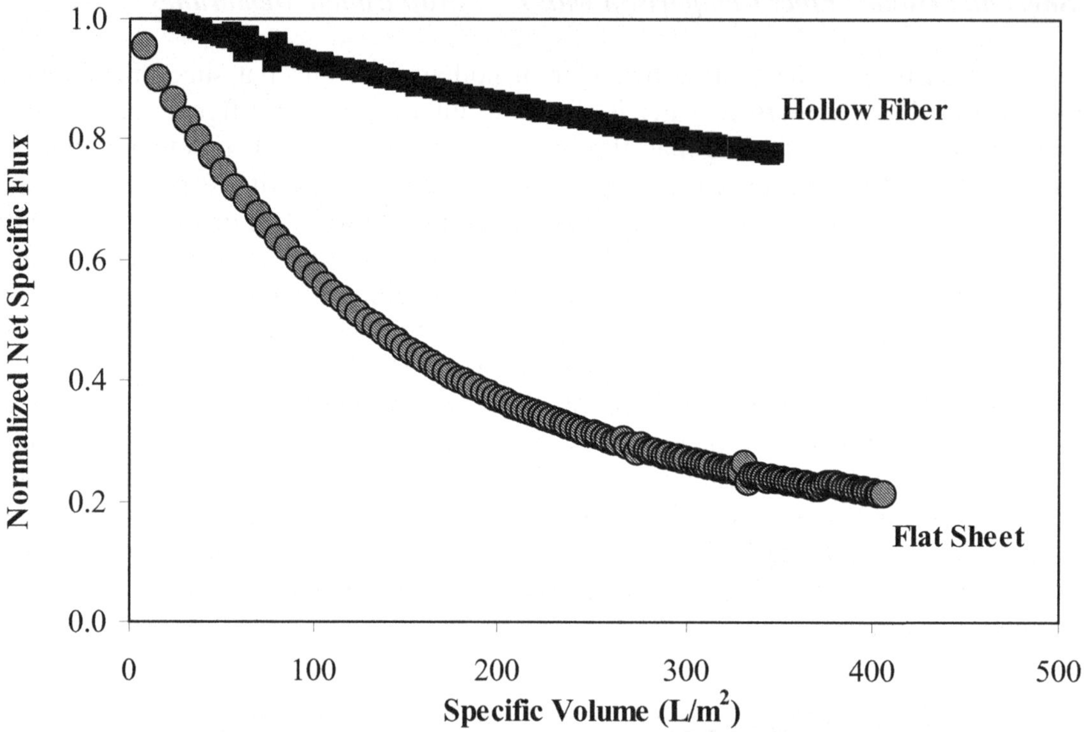

Figure 3.56 Flux of raw Waikato River water through flat sheet and hollow fiber PS ultrafiltration membranes

Flat Sheet and Hollow Fiber Comparison During Fractionation

A comparison between hollow fiber and flat sheet membranes filtering raw Seine River water with no fractionation and 1 μm fractionation is shown in Figure 3.57. For the flat sheet membrane (lower set of curves), a small increase in flux was observed after fractionation with the 1 μm membranes, as was noted earlier in this chapter. The same result is observed with the hollow fiber membranes (upper set of curves); a small increase in flux was observed after fractionation with the 1-μm membranes. These results show that the flat sheet and hollow fiber experiments demonstrate the same trend for PS ultrafiltration membrane. These also suggest that 1 μm fractionation was able to remove only small portion of the foulants in the Seine River raw water.

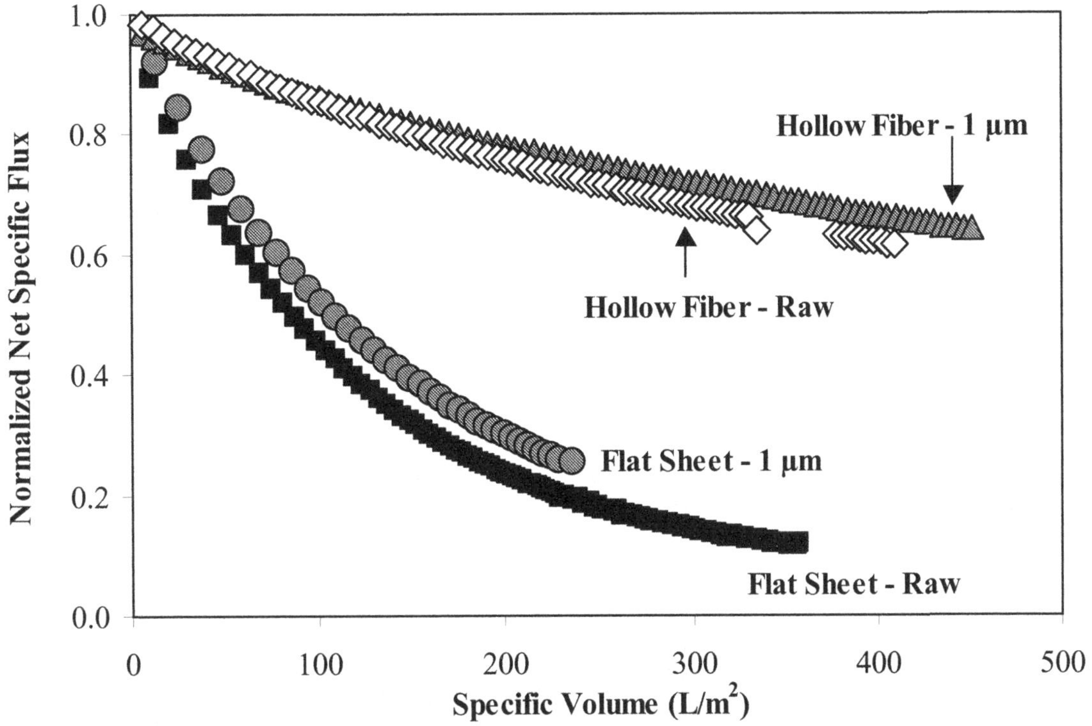

Figure 3.57 Comparison of hollow fiber and flat sheet PS membranes during fractionation of raw Seine River water

Examination of Longer-Term Fouling Trends with Hollow Fiber Membranes

Because substantial differences were noted between flat sheet and hollow fiber membranes, an additional bench-scale membrane filter configuration was developed to further investigate the discrepancy. Bench scale membrane testing, particularly with flat sheet membranes, evaluates fouling over a single filtration cycle, typically using new membranes for each test. Pilot- and full-scale plants operate over many filtration cycles, and the rate of fouling during a single filter run is of less importance than the longer-term fouling trends. A test configuration was developed to allow a small, bench-scale module to be subjected to repeated filter and backwash cycles. Feed water and backwash water were stored in separate pressure vessels, each pressurized by a nitrogen cylinder. Pinch valves in the supply tubing, operated by timer, controlled the frequency and duration of the backwash cycle. Using single-fiber membrane modules, a test with 20 to 25 filtration and backwash cycles could be completed with as little as 4 L (1.06 gal) of water. This type of testing is called long-term testing in this report.

A long-term test conducted with PVdF membranes shows the fouling effect of the raw Clackamas River water (Figure 3.58). Each single flux curve is a filter run followed by a backwash cycle. It is evident from the graph that the flux curve is steeper during a single filter run, a higher rate of fouling, than the longer-term trend established by a series of filter runs. The fouling during a single filter run can be at least partially attributed to the collection of particulate matter that is removed by backwashing, and the long-term fouling trend may be attributed more to materials that adsorb to the membrane and cannot be removed by backwashing. As a result, more than one fouling mechanism can be examined with long-term bench scale testing.

The long-term fouling of the PVdF membrane during filtration of ACH-clarified Clackamas River water is shown in Figure 3.59. The utility of this type of testing is demonstrated when these results are compared to the results with single-run flat sheet membranes presented earlier in this chapter. The ACH-clarified water shows an improvement in flux when compared to the raw water long-term test, which is consistent with the pilot scale data, but the opposite of the trend was observed with the flat sheet membranes (last graph of Figure 3.20.)

Figure 3.60 shows raw water long-term tests to compare the performance of hydrophobic and hydrophilic PVdF membranes from US filter. The hydrophobic membrane fouled more than the hydrophilic membranes, which is similar to the results demonstrated in pilot-scale testing of Chattahoochee River water (see next chapter). This type of testing may be useful for providing a more direct comparison between bench scale and pilot scale testing. However, more experimentation is needed before specific recommendations can be provided regarding the long-term fouling trends provided by this filter configuration.

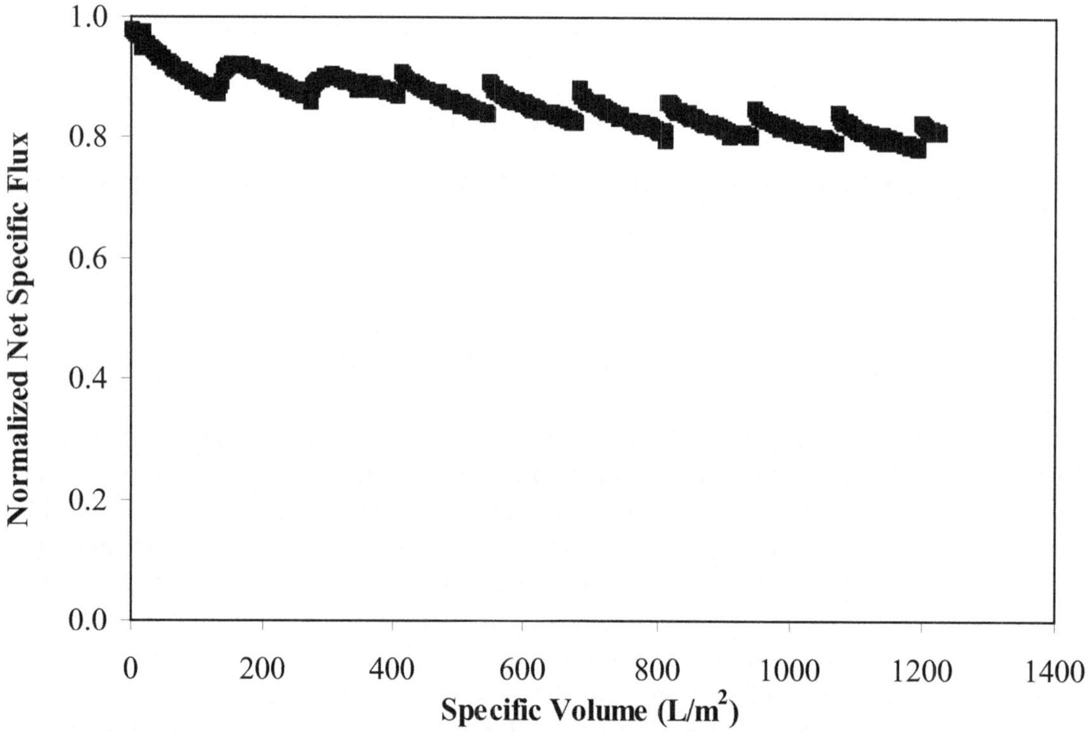

Figure 3.58 Long term test with PVdF membrane raw Clackamas River water

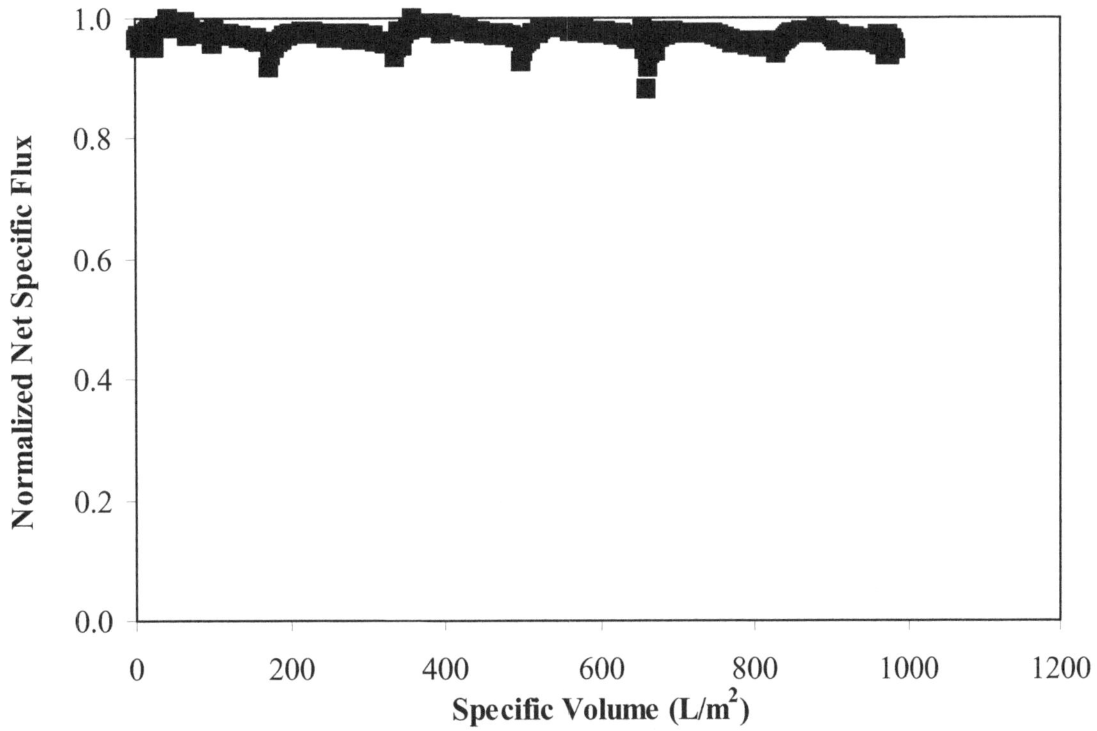

Figure 3.59 Long term test of Clackamas river water with PVdF membrane using ACH for clarification

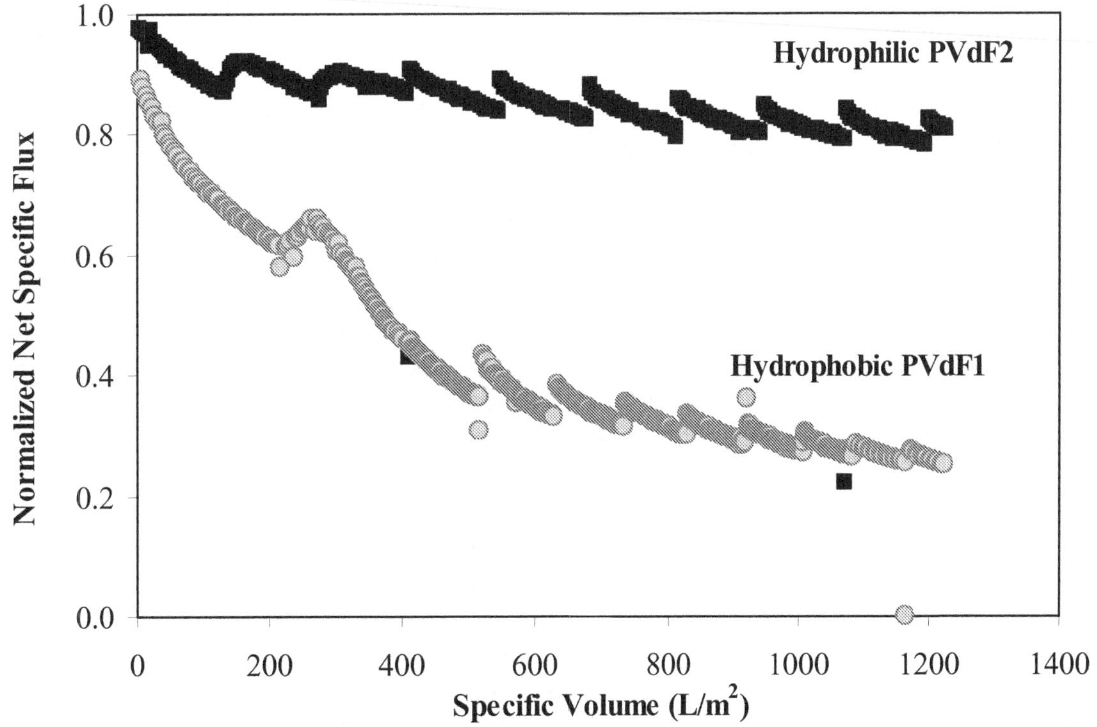

Figure 3.60 Fouling of raw Clackamas River water with PVdF hydrophobic and hydrophilic membranes

Summary of Flat Sheet and Hollow Fiber Membranes Comparison

The comparison between flat sheet and hollow fiber membranes can be summarized as follows:

1. In general, fouling of flat sheet membranes was more extensive than hollow fiber membranes, when both were exposed to the same feed water. The only exception to this was when Clackamas River water was filtered through PVdF membranes, which may have been due to the minimal fouling of this water with the flat sheet PVdF membranes.
2. The flat sheet and hollow fiber membranes generally experienced the same trends when exposed to different pretreatments, such as prefiltration or coagulation and clarification, which suggests that flat sheet membranes are an acceptable surrogate for hollow fiber membranes.
3. Long-term bench-scale testing provides a more direct comparison between bench-scale and pilot-scale testing.
4. Further research should be conducted to evaluate the performance difference of hollow fiber and flat sheet membranes, as well as the long-term testing.
5. Several lessons were learned with respect to using hollow fiber membranes at the bench-scale. It was necessary to try to develop a module with a membrane surface area similar to the flat sheet membranes that were tested. Initial modules used had 9 fibers, but used an excessive amount of water, so that filter runs were short compared to the flat sheet tests. Subsequent tests used only 1 fiber, which allowed longer filter runs. Additional improvements to the hollow fiber testing apparatus may further improve the comparison between bench scale and pilot scale results. For instance, it may be useful to operate the hollow fibers in a constant-flux mode instead of a constant-pressure mode. This could be accomplished by feeding the membranes with a peristaltic pump rather than a pressurized reservoir, and monitoring pressure instead of flux. Challenges associated with this approach are that peristaltic pumps tend to pulse, which may give different fouling results than the pulseless centrifugal pumps that are used at pilot or full scale. This negative effect may be eliminated through the use of pulsation dampeners between the pump and membrane. Continuous monitoring of pressure at the bench scale may pose an additional challenge to this approach. Flux is currently monitored in most laboratory setups by collecting the permeate on a balance, and recording the weight reading on a balance by a computer at timed intervals. This instrumentation is readily available in university environmental engineering laboratories. Recording of pressure would require pressure transducers and data acquisition software that are less readily available in laboratories, and would therefore be harder to implement. An additional advantage of feeding the membrane with a pump rather than a reservoir, however, is that it would be possible to operate the hollow fiber in crossflow mode with retentate recirculation to the feed tank, which is sometimes used in pilot or full scale applications.
6. The disadvantages and advantages of the flat sheet and hollow fiber experiments are listed in Table 3.12. Based on these advantages and disadvantages, flat sheet membranes might be more appropriate for general fouling studies (not related to specific plant performance) and studies where it is desirable to examine the fouled membranes with FTIR or SEM, and hollow fiber membranes might be more appropriate for studies in

which the greatest possible correspondence between bench and pilot/full scale systems are desired.

Table 3.12
Comparison of flat sheet and hollow fiber bench-scale membrane experiments

	Flat sheet	Hollow fiber
Advantage	- Equipment is easy to be built for utilities. - Wide variety of membrane materials and pore sizes available in scientific catalogs. - Greater degree of fouling makes it easier to observe trends in fouling from changes in pretreatment or operating conditions. - Membranes are easy to examine with SEM or FTIR	- Same material as pilot- or full-scale systems can be used.
Disadvantage	- Membrane fouling is sensitive to material properties, and flat sheet material will not be completely identical to pilot- or full-scale hollow fiber systems (even if fabricated by the same manufacturer), so tests may not be representative of the degree of fouling in hollow fiber systems. -	- Modules must be fabricated by hand in the laboratory. - Membrane options are limited to what can be obtained from commercial hollow fiber manufacturers. - Rate of fouling can be low, so trends from different operation or pretreatment can be difficult to identify during a short bench test. - Hard to examine by SEM or FTIR.

CHAPTER 4
PILOT-SCALE

The objective of the pilot-scale testing was to evaluate the effects of pretreatment chemicals and processes on the long-term performance of commercially available pilot-scale membrane systems from different manufacturers, employing different membrane configurations, membrane materials and source waters from the following plants.

- Atlanta Fulton County Water Treatment Plant (AFCWTP)
- Centre International de Recherche sur l'Eau et l'Environnement (CIRSEE)
- North Clackamas County Water Commission Plant (NCCWC)

In accomplishing the above objectives, the pilot-scale testing provided a link between the bench-scale testing and performance observed at full-scale membrane plants employing pretreatment. Table 4.1 summarizes the parameters investigated during pilot scale testing.

Table 4.1
List of pilot testing parameters

Pilot site	**AFCWTP**	**CIRSEE**	**NCCWC**
Site location	Georgia, USA	Le Pecq, France	Oregon, USA
Source water	Chattahoochee River	Seine River	Clackamas River
Membrane configuration	Pressurized	Pressurized	Submerged
Membrane material	Polyvinylidene fluoride (PVdF)	Cellulose acetate (CA) Polysulfone (PS)	Polyvinylidene fluoride (PVdF)
Pretreatment processes	Full-scale flocculation and sedimentation	Pulsatube® solids contact clarification	Pilot-scale flocculation and sedimentation
Pretreatment chemicals	Coagulant (Alum)	Coagulant (Ferric) Polymer	Coagulant (ACH, Alum) Chlorine PAC

PILOT TESTING SITES AND EQUIPMENT

Atlanta Fulton County Water Treatment Plant

Pilot Site

The treatment scheme at the full-scale AFCWTP consists of coagulation with alum, flocculation, sedimentation and dual media filtration. Chlorine is added pre- and post-filtration. Testing with pressure-driven US Filter pilot systems with two different PVdF membranes was conducted at this site. The pilot-scale test systems were set up to operate with four different feed waters diverted from the full-scale plant process flows listed below:

- raw Chattahoochee River water
- in-line direct coagulated water (with alum as coagulant)
- full-scale flocculated water (diverted from full-scale plant)
- settled water (diverted from full-scale plant)

Figure 4.1 presents a schematic of the full-scale treatment processes and the full-scale flows that were used for this study.

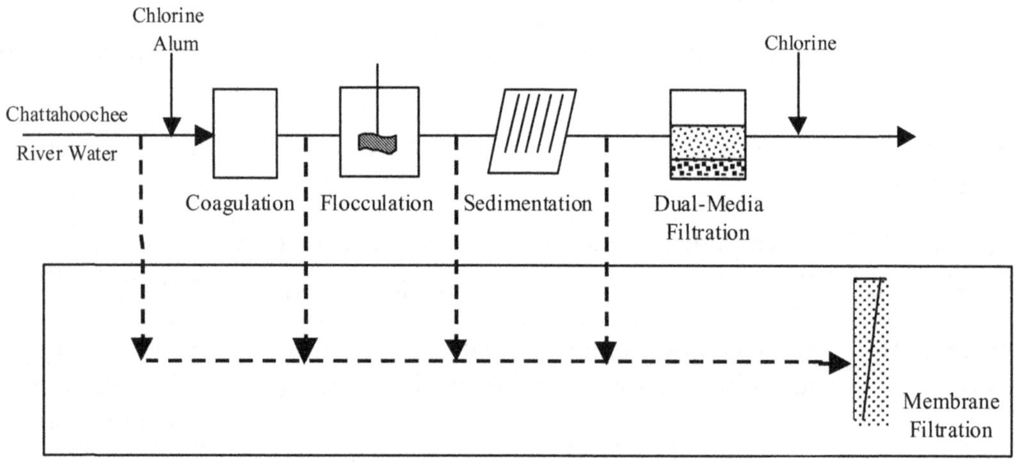

Figure 4.1 Schematic of pilot tests conducted at AFCWTP

Pilot System

The US Filter 3M10V pilot system, a pressurized dead-end membrane system, was tested at the AFCWTP test site. Two PVdF membrane modules were tested on this pilot system. The relevant specifications of the pilot systems and membrane modules are presented in Table 4.2.

Table 4.2
Specifications of the membrane system tested at AFCWTP

Parameter	PVdF1	PVdF2
Membrane Type	Hollow Fiber	Hollow Fiber
Membrane Material	PVdF	PVdF
Flow Direction	Outside-In	Outside-In
Operating Mode	Pressurized	Pressurized
Active Module Surface Area	309 ft^2 (29.7 m^2)	252 ft^2 (23.4 m^2)
Maximum TMP	22 psi (1.5 bar)	21.7 psi (1.48 bar)
Fiber Inside Diameter (mm)	0.31	0.5
Fiber Outside Diameter (mm)	0.65	0.8
Total Number of Fibers (approx.)	14,500	9,600
Active Fiber Length	3.3 ft (1 m)	3.3 ft (1 m)
Nominal Pore Size (um)	0.1	0.1
Molecular Weight Cut-off (μm)	0.1	0.03
Surface Characteristics	Hydrophobic	Hydrophilic
Surface Charge	Neutral	Neutral

North Clackamas County Water Commission Plant

Pilot Site

The treatment scheme at the NCCWC full-scale plant consists of screening followed by slow sand-filtration and chlorination. The pilot plant was set up to draw water from the raw water intake of the full-scale plant as the membrane feed water. Flocculation and sedimentation (tube settler) pilot units fabricated by MWH Research Center and Fabrication Facility were provided to investigate the membrane performance using direct and clarified water. The feedwaters that were available to the test membrane system include:

- raw Clackamas River water
- in-line direct coagulated water (with ACH as coagulant)
- flocculated water (with ACH as coagulant)
- clarified water (with ACH and alum as coagulant)
- pre-chlorinated water
- raw water with PAC

Figure 4.2 presents a schematic of the pilot-scale treatment processes and the flows that were used by the test membrane system.

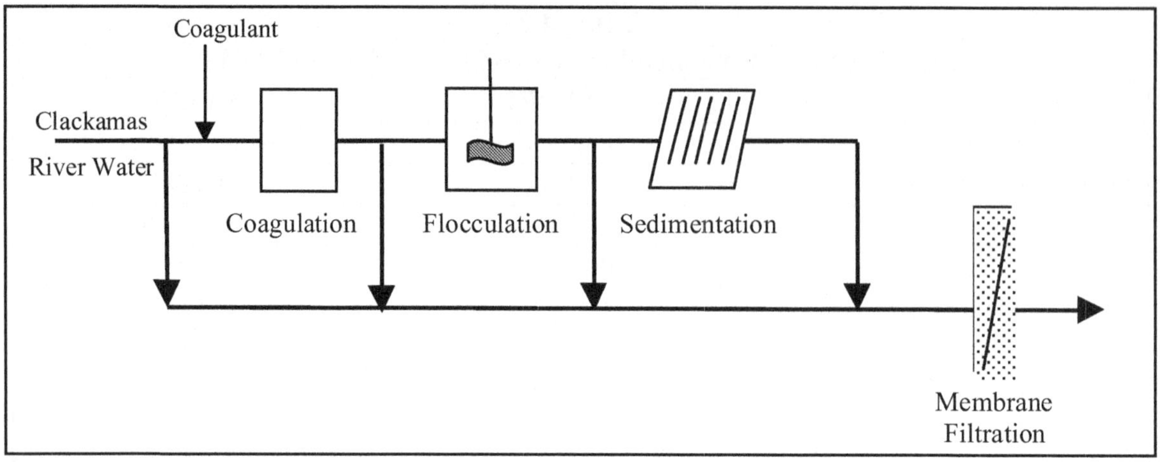

Figure 4.2 Schematic of pilot tests conducted at NCCWC

Pilot System

A submerged PVdF membrane system, US Filter S10V, was used at NCCWC plant site. The system included a 500 µm inlet strainer and air compressor for backwash. Specifications of the membrane pilot system used at this site are listed in Table 4.3.

Table 4.3
Specifications of the membrane system tested at NCCWC

Parameter	**PVdF2**
Membrane Type	Hollow Fiber
Membrane Material	PVdF
Flow Direction	Outside-In
Operating Mode	Submerged
Active Module Surface Area	272.3 ft^2 (25.3 m^2)
Maximum TMP	12.3 psi (85 kPa)
Fiber Inside Diameter (mm)	0.5
Fiber Outside Diameter (mm)	0.8
Total Number of Fibers per Module	9600
Active Fiber Length	3.4 ft (1 m)
Nominal Pore Size (µm)	0.1
Operating pH range	2-10
Maximum Temperature, °C	40

Centre International de Recherche sur l'Eau et l'Environnement

Pilot Site

Pilot testing at the CIRSEE test site consisted of a sludge blanket clarifier (Pulsator) and membrane filtration unit (Figure 4.3). The feedwaters available to the test membrane system included:

- raw Seine River water
- in-line direct coagulated water (with ferric chloride as coagulant)
- clarified water
- clarified water with polymer

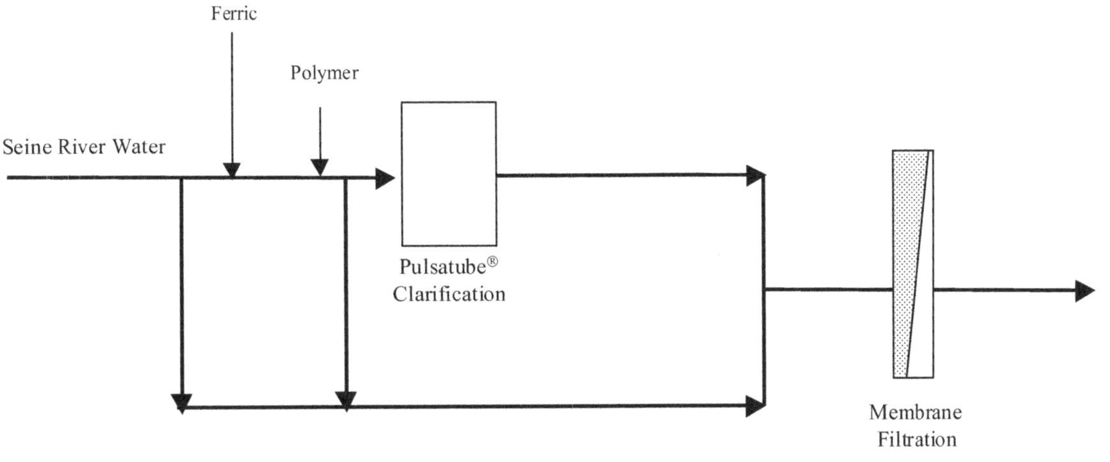

Figure 4.3 Schematic of pilot tests conducted at CIRSEE

Pilot System

An Aquasource membrane system was utilized at the CIRSEE site. The relevant specifications of the cellulose acetate and polysulfone membrane systems are presented in Table 4.4. CA membranes offer the advantage of higher flux and lower fouling potential while polysulfone membranes offer several distinct operational advantages including a higher pH tolerance, higher resistance to oxidants, and higher flexibility in cleaning.

Table 4.4
Specifications of the membrane system tested at CIRSEE

Parameter	**CA**	**PS**
Membrane Type	Hollow Fiber	Hollow Fiber
Membrane Material	Cellulose acetate	Polysulfone
Flow Direction	Inside-Out	Inside-Out
Operating Mode	Pressurized	Pressurized
Active Module Surface Area	77.5 ft^2 (7.2 m^2)	77.5 ft^2 (7.2 m^2)
Maximum TMP	20 psi (1.4 bar)	20 psi (1.4 bar)
Fiber Inside Diameter (mm)	0.93	0.96
Fiber Outside Diameter (mm)	1.67	1.65
Total Number of Fibers (approx.)	2060	2000
Active Fiber Length	3.9 ft (1.2 m)	3.9 ft (1.2 m)

(continued)

Table 4.4 (Continued)

Parameter	CA	PS
Nominal Pore Size (um)	0.01	0.01
Molecular Weight Cut-off	100 kDa	40 kDa
Surface Characteristics	Hydrophilic	Hydrophobic
Operating pH range	4 - 8.5	2 - 12
Maximum Temperature, °C	35	40

SOURCE WATER

The average raw water quality of different source waters during the pilot testing periods are displayed in Table 4.5. The values were consistent with the water quality data of the bench-scale tests. Seine River water contained the highest TOC, alkalinity, and hardness. While the values of average UV absorbance were similar among these three water sources, Clackamas River water and Chattahoochee River water had low DOC, hence low SUVA values. Clackamas River water had the highest silica concentration.

Table 4.5
Average raw water quality of different water sources

Parameter	Chattahoochee River	Clackamas River	Seine River
Temperature (°C)	13.5	18	13.6
pH	6.8	8.0	7.7
Turbidity (NTU)	5.9	2	6.3
TOC (mg/L)	1.6	1.1	3.0
DOC (mg/L)	1.5	1.1	2.8
Conductivity (μs)	40	63	510
Alkalinity (mg/L $CaCO_3$)	13	29	210
Hardness (mg/L $CaCO_3$)	13	24	261
UV (cm^{-1})	0.06	0.05	0.05
SUVA (L/mg-m)	4.0	4.5	1.9
Dissolved Iron (mg/L)	0.21	0.06	0.02
Dissolved Aluminum (mg/L)	0.11	<0.05	0.03
Dissolved Manganese (mg/L)	0.11	<0.002	0.24
Silica (mg/L)	6	20	9
Suspended Solids (mg/L)	NA	5	8.5
HPC (#/mL)	64000	230	NA
Color (Hz)	NA	NA	14.5

NA: not available

DATA ANALYSIS

In addition to the parameters presented in the bench-scale section, the following parameters were used to analyze and compare the membrane performance under various water quality and operating conditions. Feedwater recovery and productivity factor were employed to the data analysis for adjusting the difference of the pilot testing caused by backwash frequency and feedwater recovery.

Feedwater recovery (FWR) represents the net water production, and it is calculated according to the following equation:

$$\mathrm{FWR} = \left(1 - \frac{\text{Volume of water wasted}}{\text{Volume of raw water}}\right) \times 100\% \tag{4.1}$$

Productivity factor is the percentage of time the membrane system is in filtration mode divided by the combined duration of filtration and backwash.

$$\mathrm{PF} = \frac{T_f}{T_f + T_b} \tag{4.2}$$

where PF = productivity factor
T_f = time in filtration (min)
T_b = time in backwash (min)

In order to compare membrane performance with different operating conditions (i.e., backwash interval and duration), the normalized net specific flux was used to standardize the temperature adjusted specific flux by counting the feedwater recovery and productivity factor in the flux calculation. The calculation of normalized net specific flux is given as follows:

$$J_{stdnet} = J_{std} \times FWR \times PF \tag{4.3}$$

where J_{stdnet} = normalized net specific flux
J_{std} = normalized specific flux
FWR = feedwater recovery
PF = productivity factor

RESULTS

Data Reproducibility

The temperature-corrected specific flux decline rate was used to monitor system performance of the pilot-scale membrane tests, operated on various water sources. Net specific flux was used to adjust for differences in feed water recoveries, backwash interval, and backwash duration. Membrane performance at each test condition was normalized to the initial specific flux of that condition.

In order to verify the reproducibility of data collected during the pilot studies, each system was first operated at a flux that would produce fouling in a reasonable amount of time (typically 1 week). After operating to a specific flux decline of 50-70 percent, the membrane was cleaned once, according to the standard manufacturer cleaning procedure, and the specific flux recovery was calculated. At this point, if the single chemical cleaning was sufficient to restore the permeability of the membrane to its initial value, an additional cycle was performed using the same condition to demonstrate the reproducibility of results. Otherwise, successive chemical cleanings were performed to restore the permeability to its initial value before proceeding with an additional cycle.

Figures 4.4 to 4.7 show the performance of the membrane pilot plants operated on Chattahoochee (PVdF1 and PVdF2), Clackamas (PVdF2), and Seine (CA) river waters. For each membrane pilot, one standard chemical clean in place (CIP) recommended by the manufacturers was performed and the specific flux was restored to within 95 percent of its initial value for all pilot systems. No additional cleaning was required and this demonstrated the effectiveness of the standard cleaning procedure.

The membrane performance for the second cycle of operation was nearly identical to the first cycle with respect to rate of specific flux decline and run time for Chattahoochee (PVdF2), Clackamas (PVdF2), and Seine (CA) river waters, as seen in Figure 4.5, Figure 4.6, and Figure 4.7. However, the specific flux of the second cycle declined slightly faster than the first cycle in the Chattahoochee (PVdF1) pilot system (Figure 4.4); the normalized net specific flux after 2500 L/m^2 of operation was 0.5 compared to 0.6, observed in the first cycle. After investigation of the detailed data, it was found that a different water quality (increased turbidity, TOC, etc.) was observed during operation of the second cycle, compared to the first cycle, for the Chattahoochee (PVdF1) system and was most likely responsible for the altered specific flux decline rate.

Based on the data above, it was concluded that given the operating conditions (flux, backwash, FWR, etc.) and usage of the standard chemical cleaning procedure, the data collected during each phase of testing (i.e., different pretreatment condition) would be representative, comparable, and reproducible for the same membrane throughout the testing, assuming the feed water quality was stable.

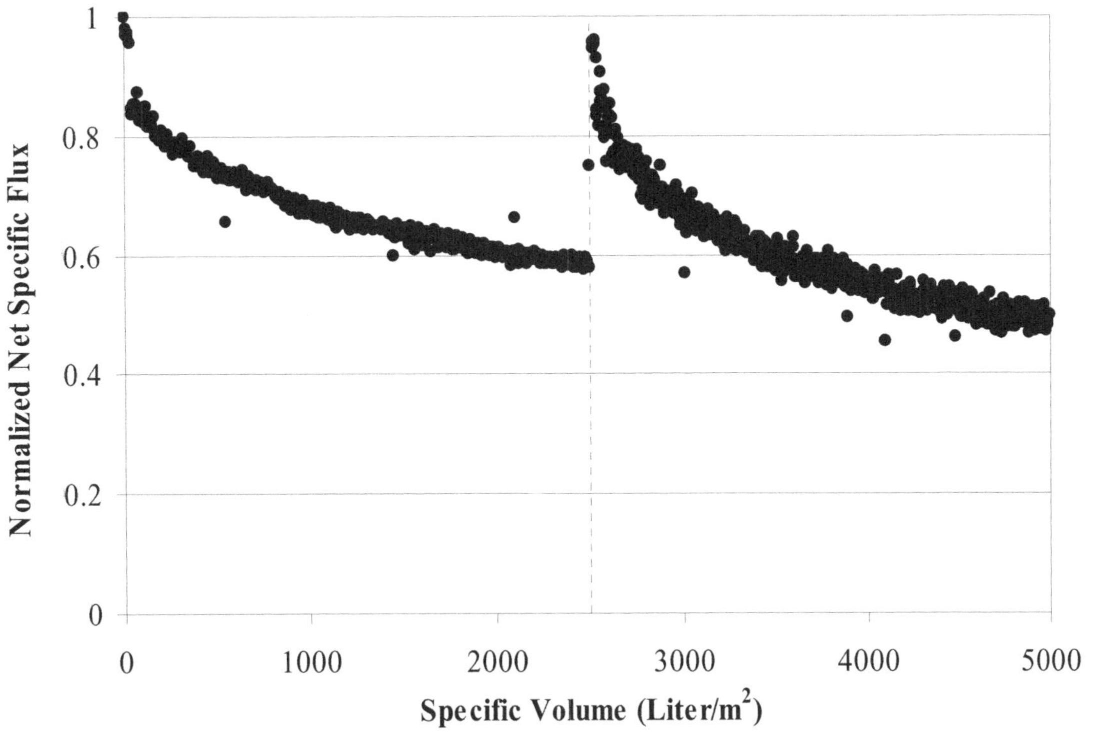

Figure 4.4 Raw water reproducibility (Chattahoochee PVdF1)

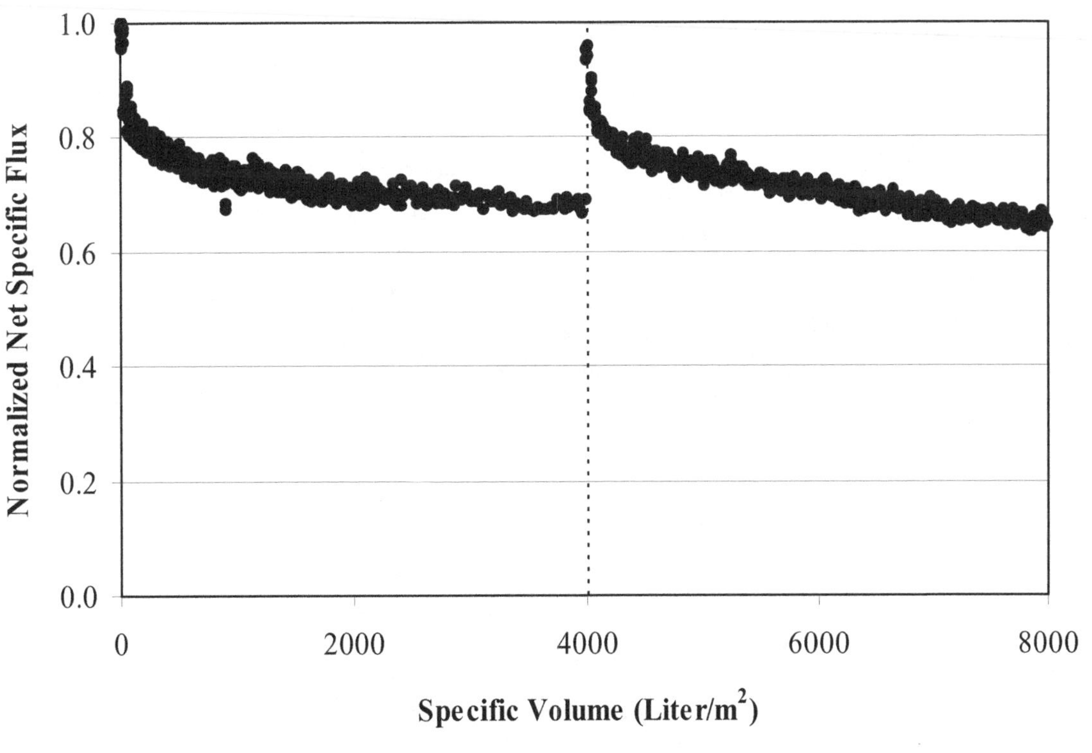

Figure 4.5 Raw water reproducibility (Chattahoochee PVdF2)

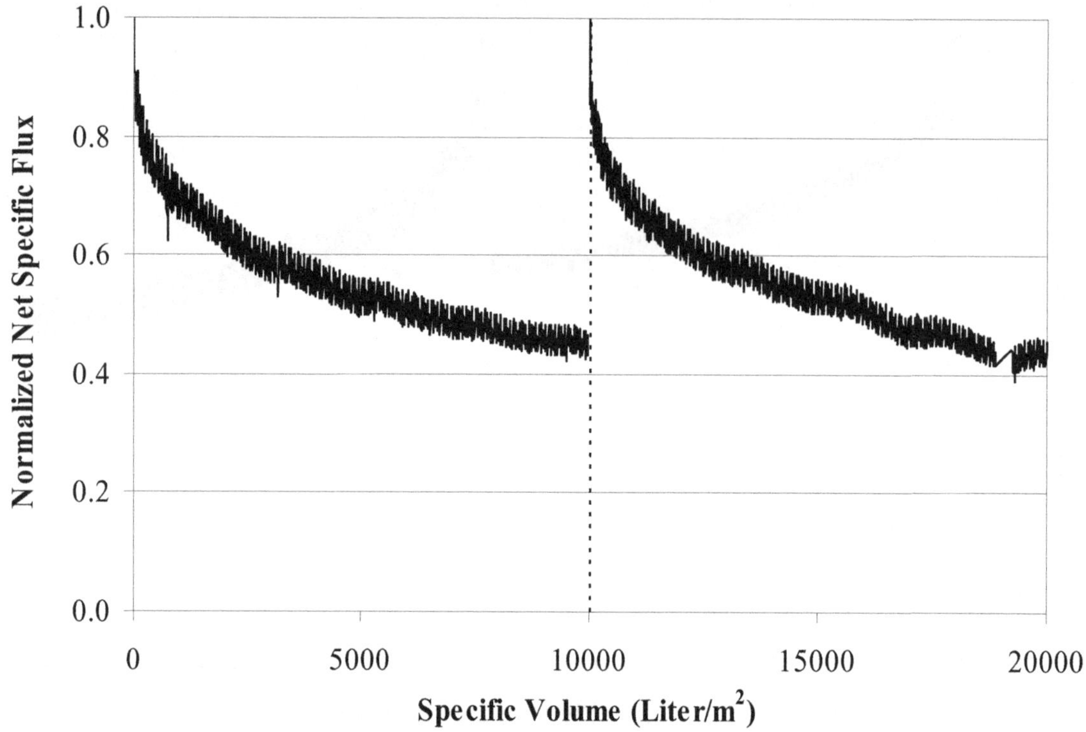

Figure 4.6 Raw water reproducibility (Clackamas River PVdF2)

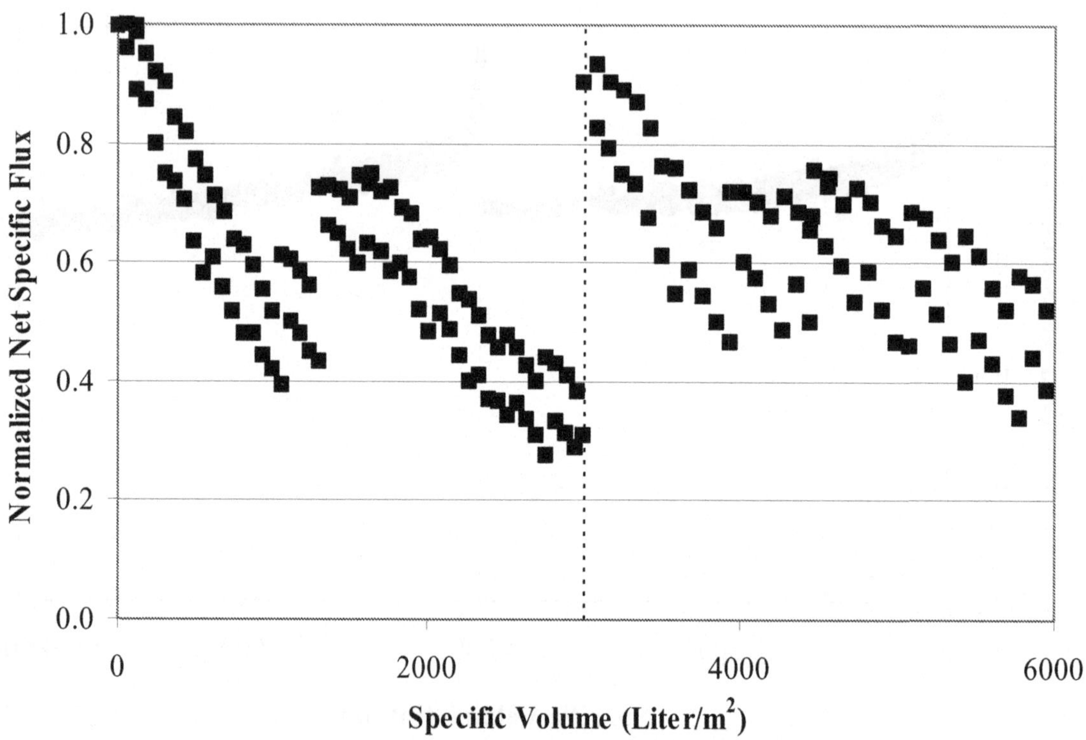

Figure 4.7 Raw water reproducibility (Seine River CA)

Effect of Pretreatment with Coagulants

Bench-scale testing demonstrated that clarified water can improve overall membrane performance by removing a portion of the foulants that would otherwise be present in raw water (i.e., TOC, turbidity, etc.). It is important to note, however, that bench-scale conclusions were derived from information collected during a single filtration cycle and may not be representative of long-term fouling. As a result, pilot-scale testing focused on evaluating the impact of coagulants on long-term membrane performance, which included regular backwash, air scour cleaning, and maintenance chemical cleaning (if necessary). The following sections summarize pilot-scale results with respect to:

- Effect of clarification
- Effect of coagulation and flocculation
- Effect of coagulant dose
- Effect of coagulant type
- Effect of polymer

Effect of Clarification

As an initial test, pilot-scale membrane plants were operated using clarified water, and compared to membrane performance on raw water alone. Figure 4.8, Figure 4.9, and Figure 4.10 compare the normalized specific flux decline of raw water and clarified water for pilots operated on Chattahoochee (PVdF1), Seine (CA), and Clackamas River (PVdF2) waters. These results demonstrate that the use of clarified water feed improved membrane performance compared to raw water. A similar trend was observed during bench-scale testing as seen in Figure 3.19, the first graph of Figure 3.20, and Figure 3.59 suggesting that the bench-scale results could qualitatively predict trends.

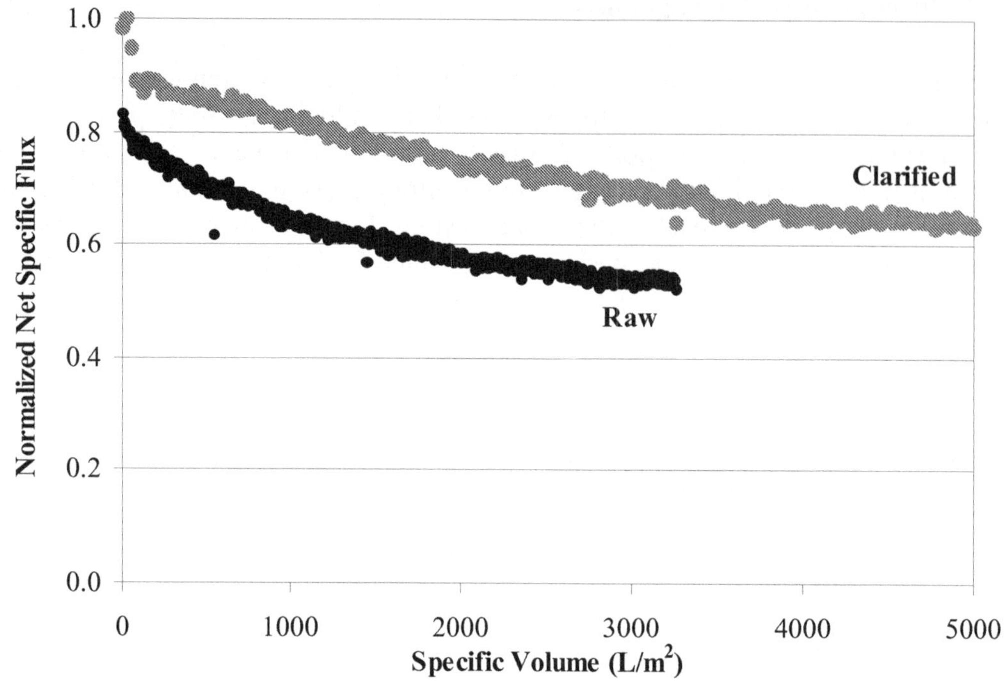

Figure 4.8 Comparison of raw versus clarified water flux decline at AFCWTP (Chattahoochee PVdF1)

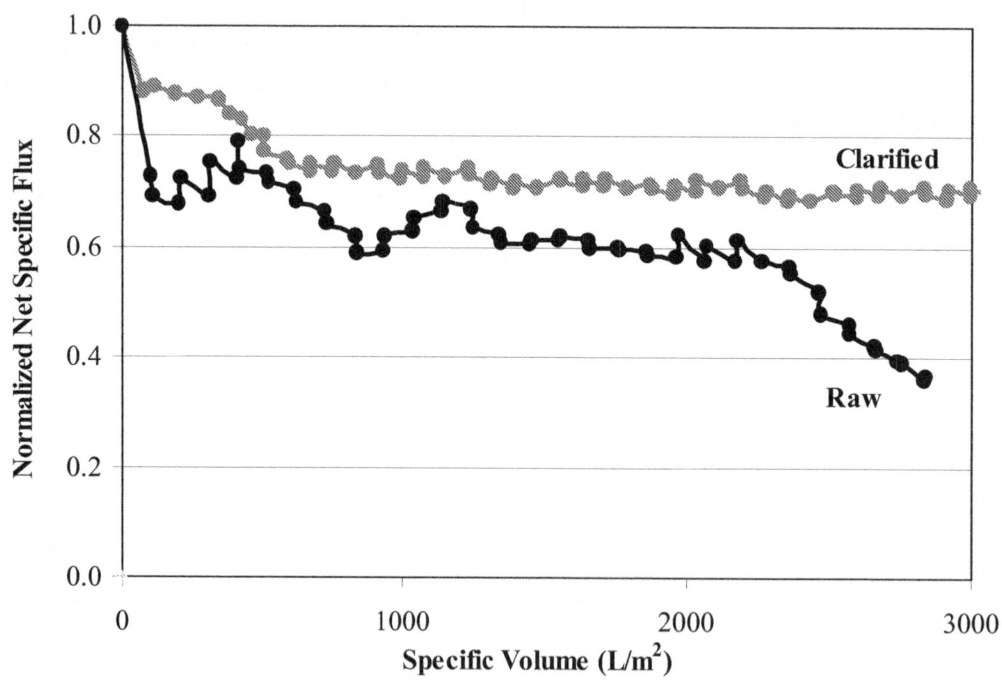

Figure 4.9 Comparison of raw versus clarified water flux decline at CIRSEE (Seine CA)

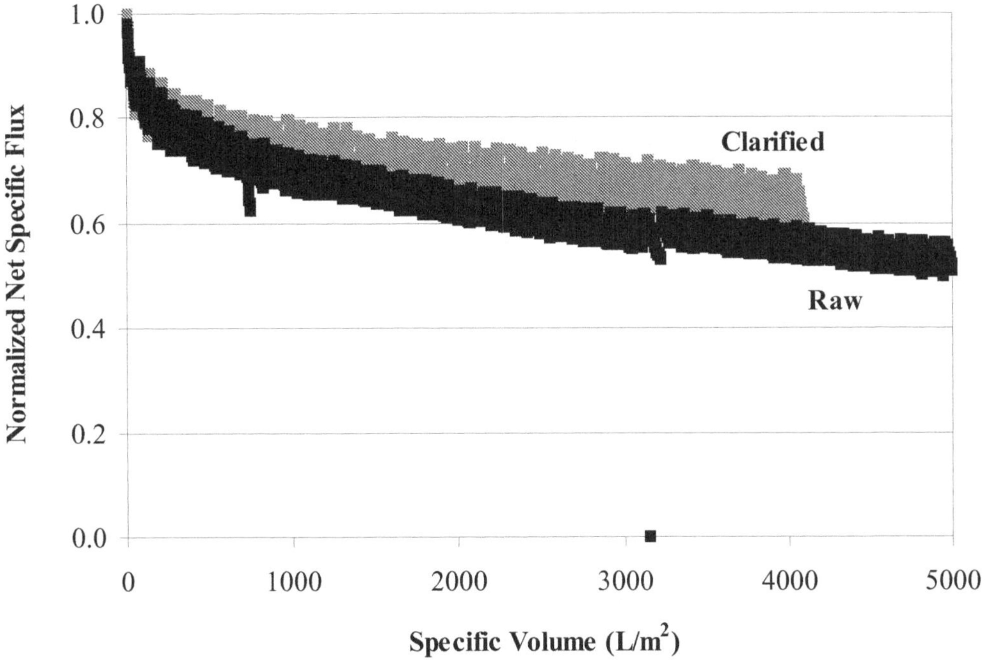

Figure 4.10 Comparison of raw versus clarified water flux decline at NCCWC (Clackamas PVdF2)

Effect of Coagulation and Flocculation

Pilot tests has also been conducted to evaluate the impacts of coagulant addition at different stages of pretreatment processes. The specific flux was monitored when the membranes were fed with inline direct coagulated water, flocculated water, and clarified water. Figure 4.11 to Figure 4.14 show specific flux decline for pilot plants operated on Chattahoochee (PVdF1 and PVdF2), Clackamas (PVdF2), and Seine River (CA) waters, respectively. The common trend noted in all figures was that direct coagulated water caused the most membrane fouling, compared to flocculated and clarified feed waters. This trend was observed regardless of membrane types (PVdF1, PVdF2, and PS). Less membrane fouling was observed with flocculated water compared to coagulated water for Chattahoochee (PVdF1 and PVdF2), and Clackamas River (PVdF2) waters. Flocculated water was not evaluated in the Seine River (CA) water pilot tests. As expected, clarified water was the best performing pretreatment condition over flocculated and coagulated water feeds. A similar trend was observed in bench-scale tests using Chattahoochee River water, in which clarified water caused less fouling than flocculated water, as shown in Figure 3.22.

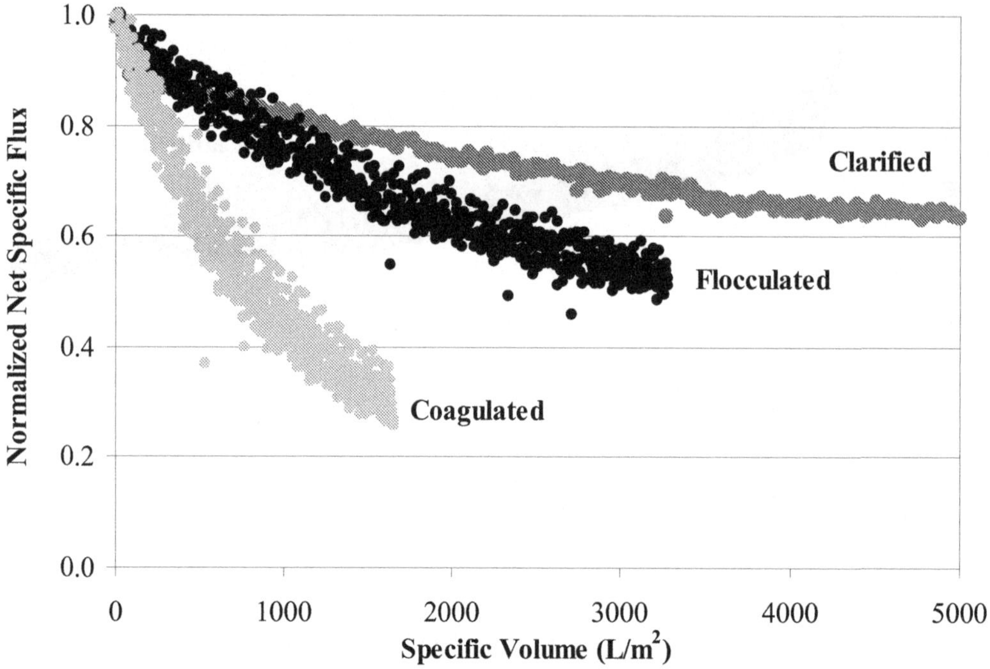

Figure 4.11 Comparison of coagulated, flocculated, and clarified water flux decline (Chattahoochee PVdF1)

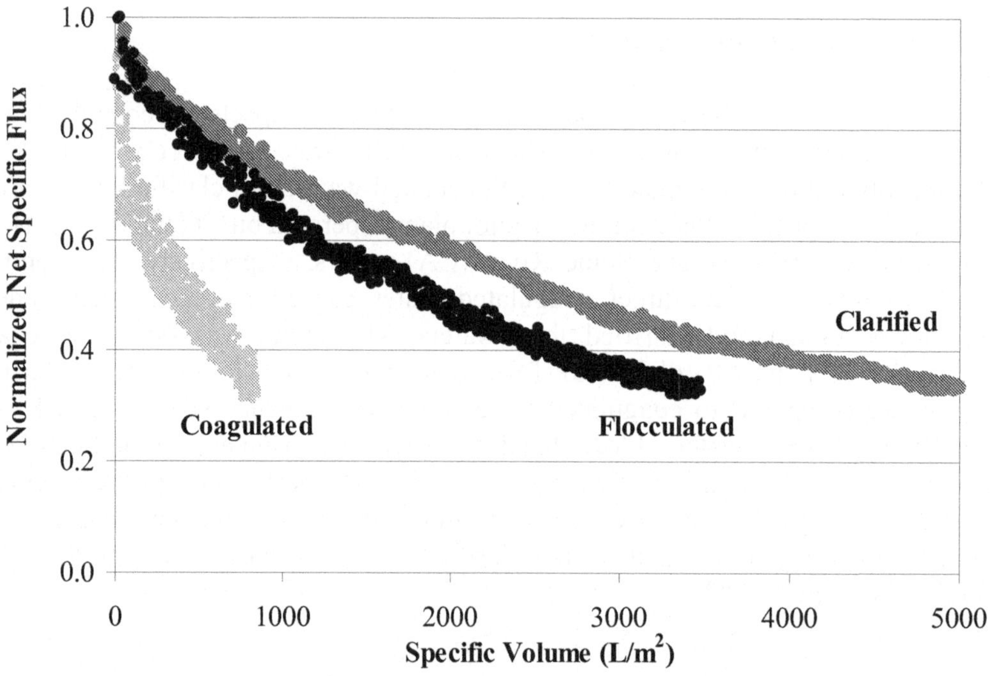

Figure 4.12 Comparison of coagulated, flocculated, and clarified water flux decline (Chattahoochee PVdF2)

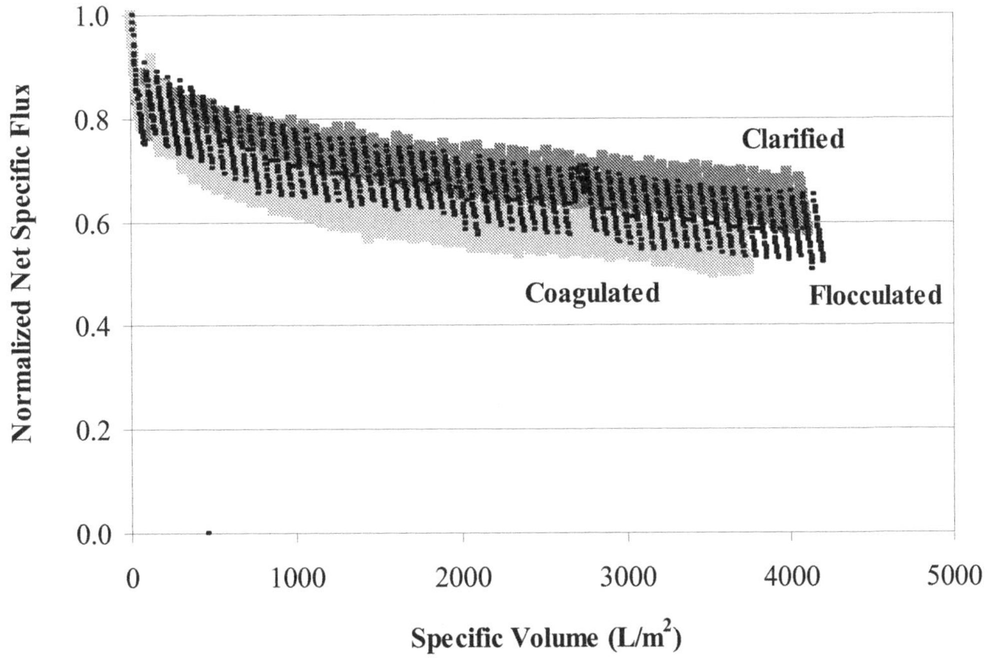

Figure 4.13 Comparison of coagulated, flocculated, and clarified water flux decline (Clackamas PVdF2)

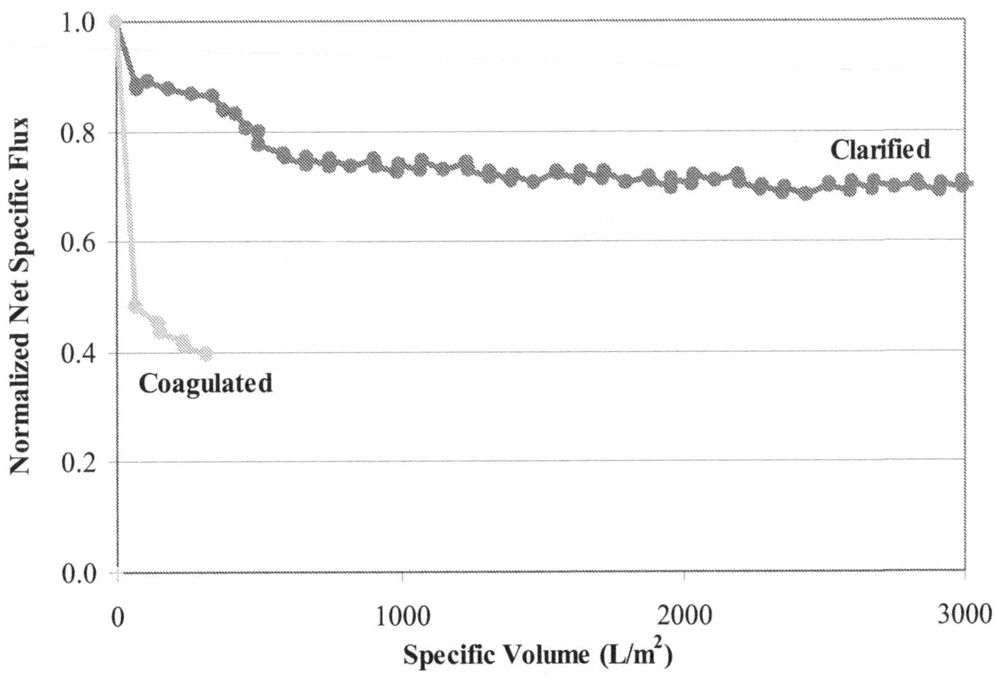

Figure 4.14 Comparison of coagulated, flocculated, and clarified water flux decline (Seine CA)

Effect of Coagulant Dose

To obtain a better understanding of the impact of coagulant dose, pilot-scale tests with Clackamas River (PVdF2) water were performed by varying the coagulant (ACH) dose. Figure 4.15 shows the observed membrane performance when two different coagulant doses were used to generate clarified water. For the majority of pilot testing, a low coagulant dose (0.2 mg/L as Al) was used for preparing coagulated, flocculated, and clarified feed water to the membrane. This chosen dose was based upon previous pilot-scale membrane testing at the NCCWC plant and was assumed to be an optimized dose. After completing the PVdF2 membrane tests with clarified water (and 0.2 mg-Al/L), the membrane was cleaned according to standard manufacturer cleaning procedures to restore the membrane to its initial permeability. The coagulant dose was then increased to 0.5 mg-Al/L and the clarified water was fed to the membrane pilot. The 0.5 mg-Al/L dose was chosen based on the ACH dose used during bench-scale testing which was determined through jar test.

According to Figure 4.15, membrane performance with 0.5 mg-Al/L ACH deteriorated compared to that fed with raw water. This was consistent with similar results observed during bench-scale testing (Figure 3.20 – Clackamas River waters), in which clarified water (at the same metal mole dose of coagulant, 0.5 mg/L) fouled more rapidly than raw water. The unpredictable fouling appeared to be due to the presence of a fouling species created by the interaction of high levels of silica and aluminum-based coagulants. Moreover, these results demonstrated that a small change in the coagulant dose could have a large impact on the membrane performance – an increase of only 0.3 mg/L as Al caused the PVdF2 membrane to foul rapidly.

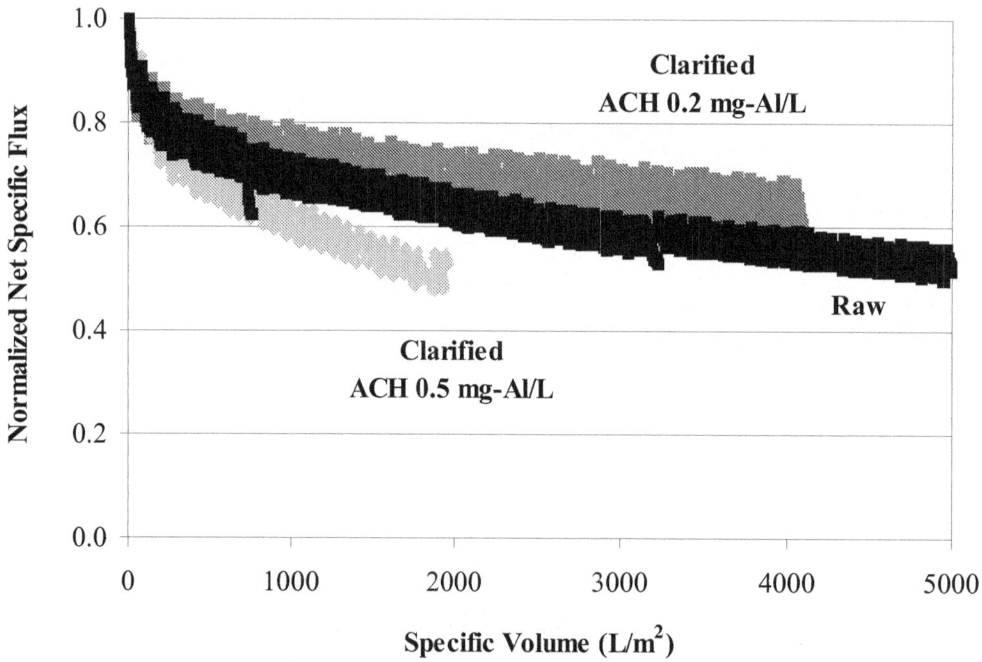

Figure 4.15 Impact of coagulant dose on membrane performance with clarified water (Clackamas PVdF2)

The relationship of decreasing performance with increasing dose, however, cannot be generalized, as an opposite trend (increased performance with increasing dose) was observed during bench-scale testing with CA membranes on Medina River water (Figure 3.34). These bench- and pilot-scale findings clearly indicate the effect of coagulant on membrane performance cannot be fully predicted because the coagulant type and coagulant dose are highly dependent on site-specific water quality parameters. Coagulant doses, typically optimized based on traditional water quality parameters from conventional jar testing (i.e., turbidity, TOC, SUVA, etc.), may be used as the preliminary selected dose, and need to be verified on site for membrane performance. This statement is also supported by the results of bench-scale testing that no single measured water parameter (e.g., alkalinity, hardness, TOC, SUVA, or silica) could be concluded as the predominant cause of fouling. As a result, it is recommended that if membrane pretreatment using coagulants is desired, a pilot-scale investigation be performed to optimize the coagulant type and dose for membrane performance.

Effect of Coagulant Type

Further investigation of the impact of coagulant type was performed using Clackamas River water and PVdF2 membranes. Figure 4.16 shows the performance of membranes fed with ACH and alum clarified waters. Pilot-scale results were consistent with bench-scale tests performed using PVdF membranes on Clackamas River water, in which membrane performance improved when fed ACH at the same metal mole equivalent dose as alum. Additionally, a similar trend was observed during bench-scale testing with PS membranes (Figure 3.28), where membrane performance improved when fed ACH clarified water over alum and ferric coagulants. These results clearly suggest that ACH may be an ideal coagulant for membrane pretreatment, independent of membrane type. However, as stated in the previous section, care must be taken to ensure that the coagulant dose used for membrane pretreatment be optimized based on site-specific water quality.

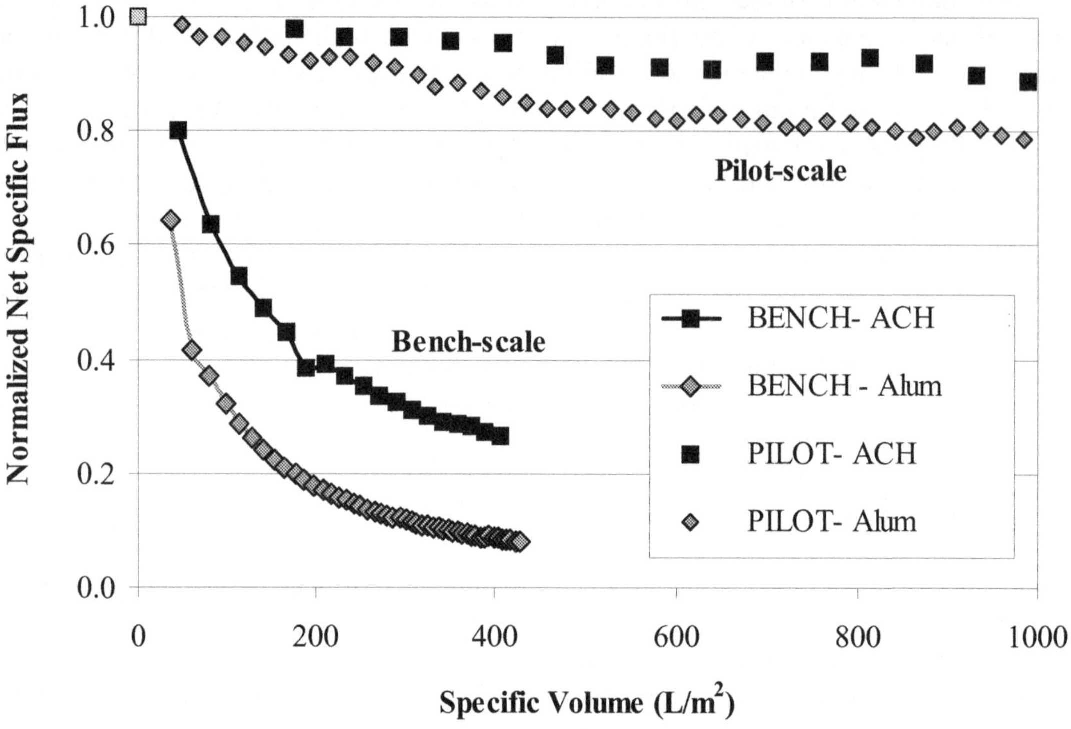

Figure 4.16 Bench- and pilot-scale membrane performance of clarified water using ACH and alum coagulants (Clackamas PVdF)

Effect of Polymer

For conventional treatment processes, polymers are frequently added in conjunction with coagulants to produce an improved water quality. The addition of polymer usually increases the settling velocity of the coagulated floc particles, thereby improving the clarified water quality. The pilot-scale testing at CIRSEE included the test to evaluate the impact of polymer.

In these tests, ferric chloride was used as the coagulant and Rhodopol polymer was added with a dosage of 0.08 mg/L. The Rhodopol, a heteropolysaccharide, anionic polymer based on natural gum of xanthan, was identified as the suitable polymer for the Seine River water by screening tests. Table 4.6 shows the average water quality parameters of raw and pretreated Seine River waters during the pilot testing with CA membranes. The water quality was improved in the clarified water with polymer compared with the clarified water without polymer (e.g. lower solid, turbidity, dissolved iron, and alkalinity concentration).

Figure 4.17 compares the membrane performance of raw water to clarified water with and without polymer. When the membranes were fed with raw water, the normalized specific flux dropped to 0.37 after specific volume reached 2800 L/m^2. Using ferric chloride alone, the clarification improved the membrane performance by 91 percent (normalized specific flux decline to 0.71 after specific volume of 2800 L/m^2). The membrane performance was improved even further when polymer was added to produce clarified water. As seen in Figure 4.17, membrane fouling was minimal when polymer was used, and the normalized specific flux decline improved by 143 percent (normalized specific was 0.90 after specific volume of 2800 L/m^2) compared to the raw water flux decline. The results confirmed that an improvement in membrane performance could be observed when carefully selected polymer is used with coagulant as membrane pretreatment.

Concurrent bench-scale tests were not performed to evaluate the effect of polymer addition. However, similar to choosing an appropriate coagulant dose, polymer doses should also be optimized based on site-specific membrane performance and not solely on conventional water quality indicators (e.g. turbidity) from standard jar testing.

Table 4.6
Average water quality parameters of clarified Seine River water with and without polymer

Parameter	Raw water	Settled water ($FeCl_3$)	Settled water ($FeCl_3$+Rhodopol)
pH	7.7	7.6	7.4
Turbidity (ntu)	6.3	1.5	0.9
Suspended Solids (mg/L)	8.5	3.0	1.6
UV_{254} (cm^{-1})	0.052	0.038	0.041
TOC (mg/L)	3	2.4	2.5
DOC (mg/L)	2.8	2.3	2.4
Dissolved Iron (mg/L)	0.023	0.035	0.021
Dissolved Manganese (mg/L)	0.238	0.472	0.109
Alkalinity (mg/L $CaCO_3$)	210	219	175

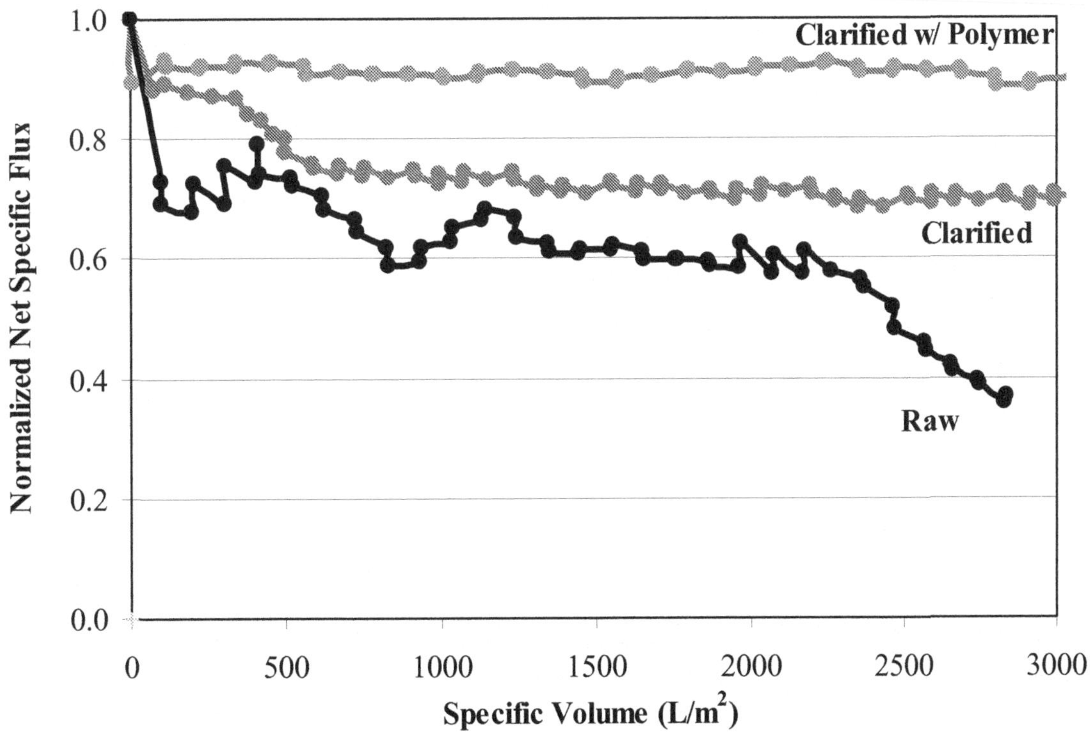

Figure 4.17 Impact of polymer addition on membrane performance (Seine CA)

Effect of Alternative Pretreatment with Chlorine

The addition of chlorine to raw water before membrane treatment has the potential benefit of oxidizing organic matter that may normally contribute to the organic fouling of membranes. Pilot-scale testing demonstrated that when Clackamas River water was pretreated with chlorine, little to no effect on PVdF2 membrane performance was observed. Similar results were demonstrated during bench-scale testing with PVdF membranes on Clackamas River water, when the same specific volume of water was filtered. Figure 4.18 compares results obtained in both bench- and pilot-scale test indicating that the membranes fouled by approximately 20 percent after 400 L/m^2 of filtration. At this filtration length, the addition of chlorine to the raw water was not beneficial to the process.

Pilot-scale testing, however, allowed the impact of chlorine addition to be evaluated during extended operation. Figure 4.19 compares the long-term normalized specific flux decline of raw and chlorine-treated Clackamas River water using PVdF2 membranes. As previously described, membrane performance was nearly identical during the initial operation. However, after approximately 1500 L/m^2 of filtration, membrane fouling slowed for the chlorinated feed. Compared to the membrane performance when fed raw water alone, the PVdF2 membrane continued to foul to approximately 0.45 after 10,000 L/m^2.

These results suggest that long-term membrane performance can be enhanced by using pre-chlorination. It is anticipated that chorine controls biofouling that would normally occur when membranes are fed raw water alone.

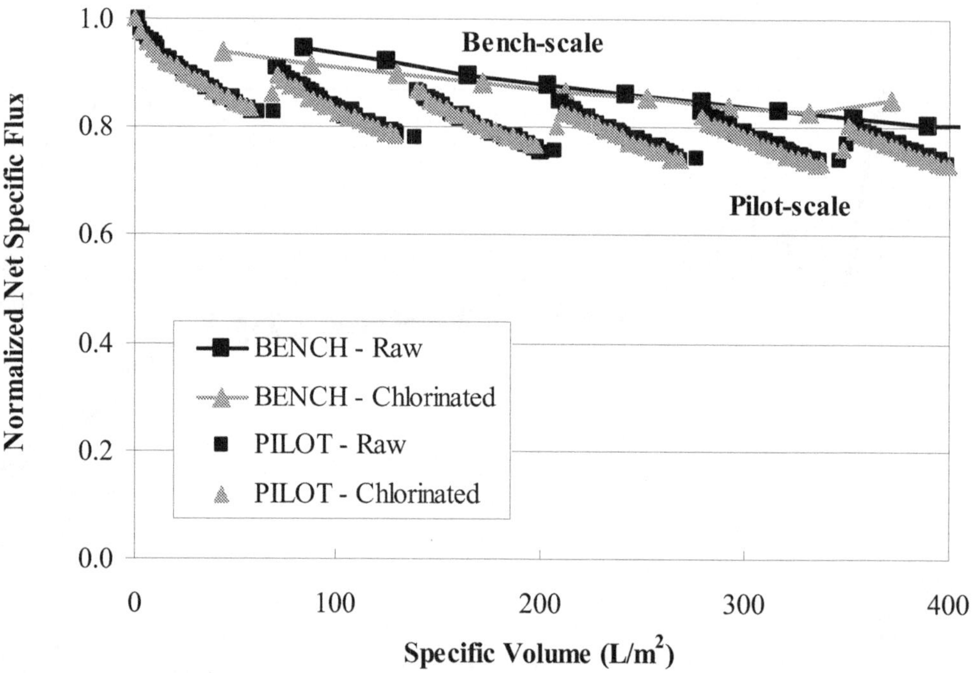

Figure 4.18 Impact of chlorine pretreatment on short-term bench- and pilot-scale membrane performance for PVdF membranes

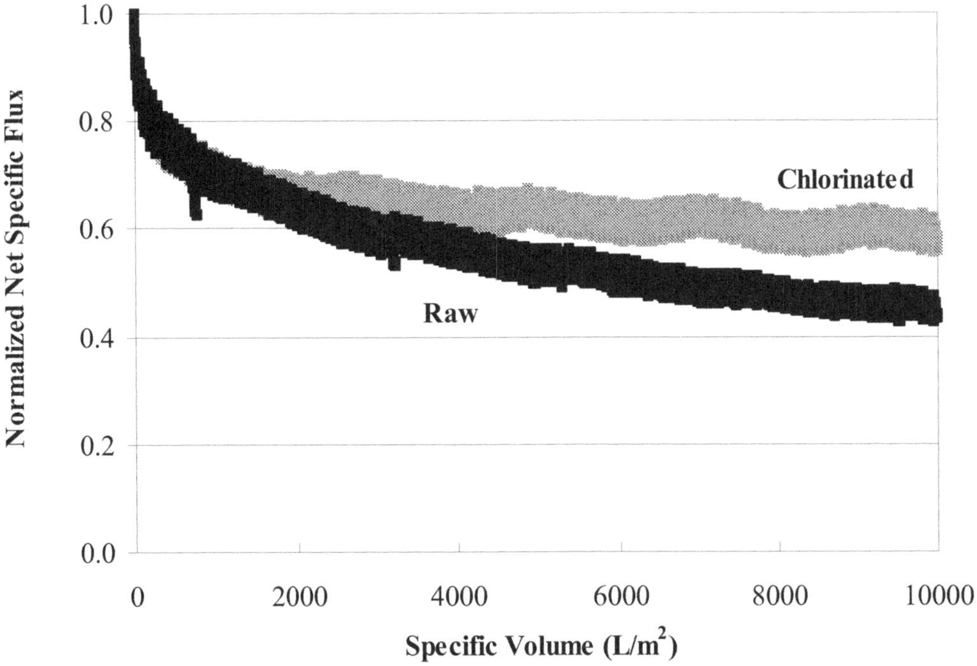

Figure 4.19 Impact of chlorine pretreatment on long-term pilot-scale membrane performance (Clackamas PVdF2)

Effect of Alternative Pretreatment with Powdered Activated Carbon

As cited in the literature review section, PAC has been commonly used for taste and odor control or organic removal as a pretreatment process before membranes. The PAC is utilized for the organic removal and the membrane is to separate the PAC and other particulate.

To investigate the effect of PAC on membrane performance, pilot-scale tests were performed using Clackamas River water from the NCCWC WTP, which has seasonal taste and odor issues during algae blooming periods. A bench-scale study suggested that 50 mg/L of PAC with 15 minutes of contact time may reduce MIB and geosmin concentrations approximately 1 log for this water. Therefore, the pilot test was conducted with the addition of 50 mg/L of PAC to the raw water. Figure 4.20 shows the comparison of normalized net specific flux decline between raw water and PAC addition for both bench- and pilot-scale results. In bench-scale results, the PAC addition had no impact on membrane fouling, but caused more fouling in pilot-scale study. It must to be pointed out that the bench-scale study of Seine River water used low PAC dose, 2 mg/L, which is an optimized dose for atrazine removal at the full-scale Vigneux plant using Seine River as source water. As discussed in the section regarding coagulant dose, the PAC dose should be optimized for site-specific water quality along with other pretreatment processes.

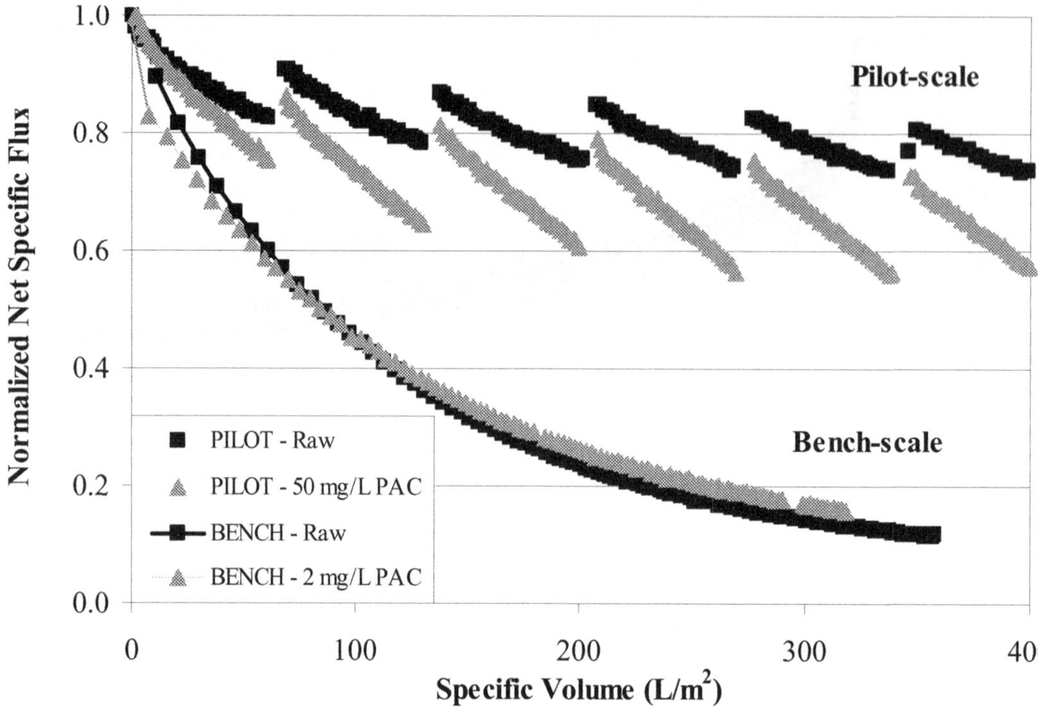

Figure 4.20 Impact of PAC as pretreatment for bench-scale (Seine PS) and pilot-scale (Clackamas PVdF2) membrane performance

Effect of Membrane Material

With several different types of membranes currently on the market, it is important to consider the selection of membrane material based on site-specific water quality. Raw water may contain constituents (e.g., high iron, silica, organic compounds, etc.) that can interfere with the membrane or pretreatment chemicals. Pilot-scale testing focused on three main types of membranes commonly used: PVdF, PS, and CA membranes. Generally speaking, PS and CA membranes have similar properties, including low to moderate chlorine tolerance and typically being classified as ultrafiltration membranes. PVdF membranes, on the other hand, are highly chlorine tolerant and typically have larger pores. Pilot testing was performed to determine impact of membrane material on membrane flux decline.

Figure 4.21 shows that the CA membrane performed slightly better than the PS membrane when operated on clarified Seine River water using ferric chloride as coagulant. Similar results were observed during bench-scale testing conducted using Medina River water (Figure 3.32). Although both CA and PS are ultrafiltration membranes, the material difference had an impact on membrane fouling.

Figure 4.22 compares the pilot-scale performance of two different PVdF membranes using Chattahoochee River water. Although the base material was the same for these two tests, slight difference in membrane surface chemistry had significant effect on membrane fouling. The PVdF1 membrane is a hydrophobic membrane compared to the more hydrophilic PVdF2 membrane. Long-term bench-scale testing (Figure 3.60) also demonstrated similar membrane performance using Clackamas River water. Both the bench- and pilot scale studies illustrated that the hydrophilic PVdF2 membrane performed better than relatively hydrophobic PVdF1 membrane.

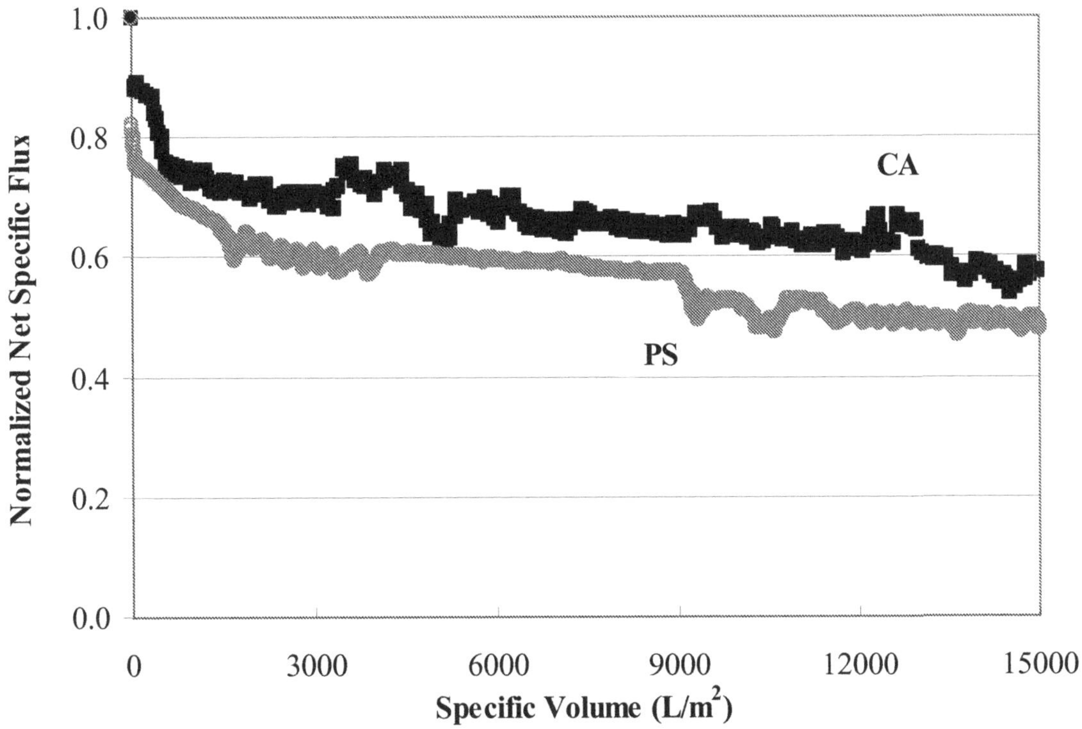

Figure 4.21 Impact of membrane material (CA versus PS) pilot-scale membrane performance with clarified Seine River water

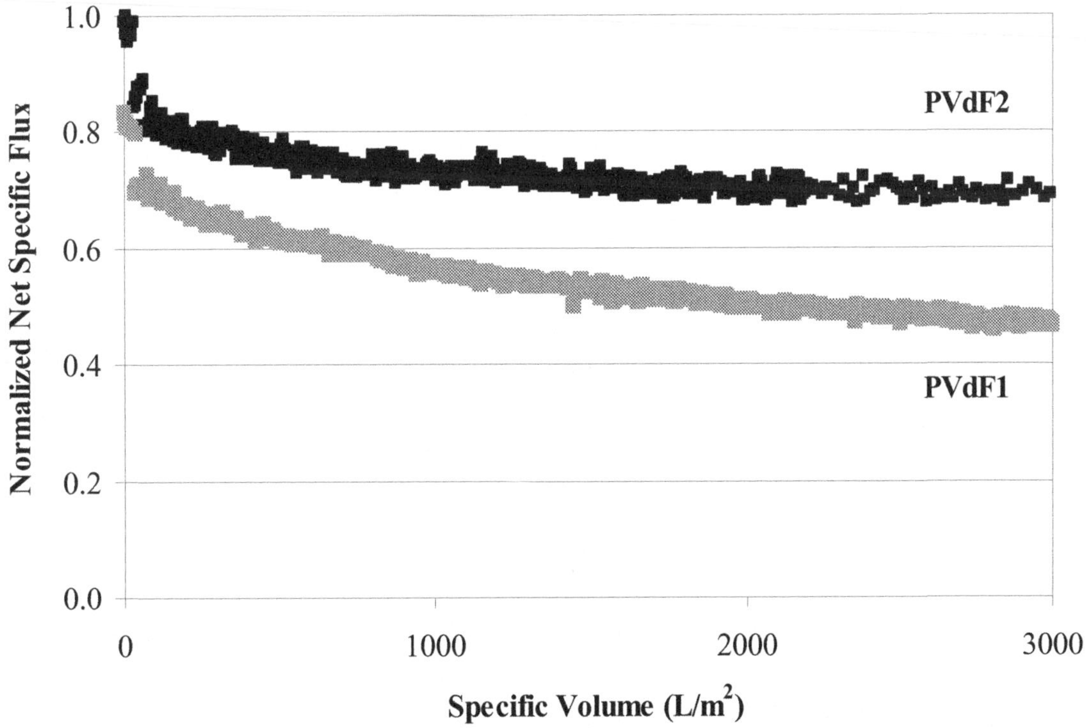

Figure 4.22 Impact of membrane material (PVdF1 versus PVdF2) on pilot-scale membrane performance (Chattahoochee River)

CHAPTER 5
FULL-SCALE

The objective of the full-scale study was to expand the bench- and pilot-scale evaluation of membrane performance impacted by pretreatment to the full-scale level. Several membrane plants with various pretreatment processes were contacted to obtain general plant information, membrane performance data, water quality data, chemical usage, and lessons learned. The full-scale results assisted in identifying the trends of the bench- and pilot-scale studies as well as the design and operational issues that need to be considered when implementing MF/UF membranes with pretreatment.

FULL SCALE PLANT DESCRIPTIONS

The selected treatment plants covered a wide range of geographic locations, pretreatment processes, membrane materials, and membrane manufacturers. Table 5.1 presents the general information of the plants that participated in the full-scale evaluation study.

Table 5.1
Full-scale plants

Utility Name	Pretreatment	Membrane Material	Membrane Configuration	Membrane Manufacturer
BexarMet, Texas, USA	PAC (optional), Ferric Sulfate, Superpulsator®	CA	Pressurized	Aquasource
Vigneux, France	Coagulation (Alum), Flocculation, Clarification, GAC, Ozonation, PAC	CA	Pressurized	Aquasource
Ashland, Wisconsin, USA	In-line Coagulation (ACH)	PP	Pressurized	US Filter
Meredith, Victoria, Australia	In-line Coagulation (ACH)	PP	Pressurized	US Filter
Westminster, Colorado, USA	Permanganate, Coagulation (PACl), Flocculation, Clarification	PVdF	Pressurized	Pall
Waikato, New Zealand	Lime, Alum, Superpulsator®	PVdF	Submerged	Zenon

RESULTS

Unlike the bench- and pilot-scale studies, which were conducted under controlled conditions, the full-scale evaluation could only rely on the operational conditions of the utilities. Therefore, direct comparison among these plants are not provided, but each plant is presented as a case study. Complete case studies of all the plants are included in Appendix B. This chapter summarizes selected case studies and presents the important findings including links to the bench- and pilot-scale studies. Lessons learned related to the pretreatment processes from these utilities are also provided.

Effect of Clarification

The results of pilot-scale section concluded that clarification improved membrane performance compared to raw water. The City of Westminster's Northwest Water Treatment Facility (Westminster) further supports pilot-scale results comparing raw water to clarified water.

Westminster located in Colorado, USA is a 15 mgd microfiltration plant with eight PVdF pressurized membrane skids. The plant has a unique setup for operating the membrane systems with or without pretreatment. During the season when the water quality is above average, the membrane system is usually operated without any pretreatment. The direct membrane filtration utilizes seven skids for operation and the remaining skid is used as a secondary membrane to treat the backwash water. When the membrane system is operated with pretreatment, backwash water is transported to the front of the plant (i.e., the backwash water is recovered by the pretreatment process).

Figure 5.1 presents the membrane specific flux versus specific volume. Membrane pretreatment was utilized to produce clarified water during phase 1 and phase 3. The membranes fouled severely after the pretreatment was ceased in phase 2, and the fouling was to the extent that even after resuming the complete pretreatment in phase 3, the specific flux could not be recovered. A clean-in-place (CIP) with citric acid was performed to restore the permeability of the membrane. The specific flux decline rate of the membrane system without pretreatment was more than twice of the specific flux decline rate when the membrane system was operated with pretreatment.

Coagulant doses were controlled to maintain settled water turbidities just below 1.5 NTU. Additional coagulant (PACl) improved settled water turbidities but did not affect membrane filtration or fouling rates. The finding that increase coagulant concentration could further reduce turbidity but not improve membrane performance supports the conclusion that an optimal coagulant dose based on typical water parameters (e.g., turbidity) monitored in jar tests may not be the ideal dose for membrane performance. A coagulant dose determined according to a jar test should be further verified through membrane performance monitoring. Details of this case study are included in Appendix B.

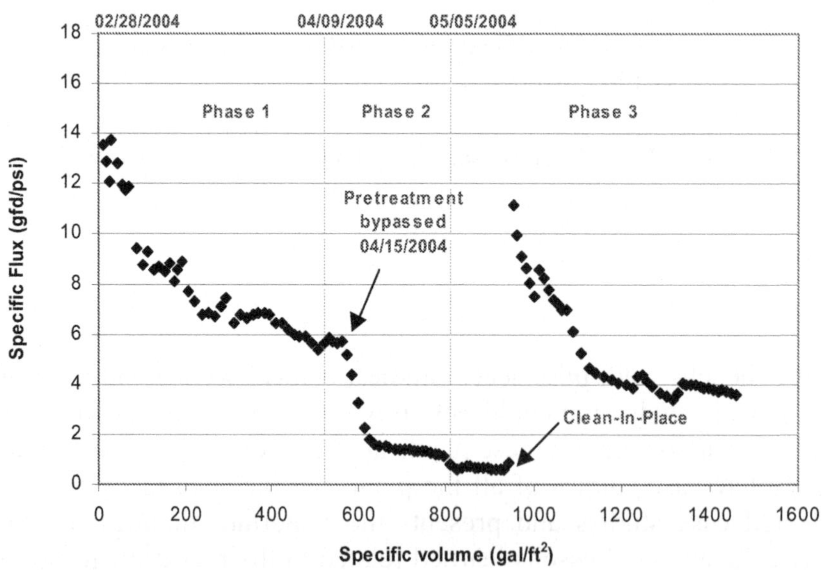

Figure 5.1 Westminster membrane performance data

Effect of Coagulation

Although clarification improved membrane performance and in-line coagulation caused more fouling than raw water during pilot-scale tests, the full-scale Ashland Microfiltration Water Treatment Plant (Ashland) showed that in-line coagulation at a low dose could assist in stabilizing the membrane TMP.

Ashland located in Wisconsin, USA is a 1.2 mgd microfiltration plant with pressurized polypropylene membranes. Raw water is directly coagulated with ACH and passed through a 500 μm strainer before entering the membrane system. Chlorine is used as disinfectant after membrane filtration. The microfiltration system includes four skids, and each skid contains 48 individual membrane modules. The flow is equally distributed to the four skids. The systems are operated with a 45 minutes filtration cycle followed by 2.5 minutes backwash. Chemical cleaning, which is applied approximately monthly, employs phosphoric acid at pH of 2, followed by ethylenediaminetetraacetic acid (EDTA), sodium hydroxide, and hydrogen peroxide.

Typically, raw water is pretreated with in-line coagulation using 0.4 mg/L ACH during the months of May and June for color removal. Based on the water quality analysis, the color was successfully removed from the maximum value of 25 true color unit (TCU) to 5 TCU with in-line coagulation and membrane filtration. However, the TOC levels between raw water and filtrate remained similar.

The ACH dosage was in the range recommended by the membrane manufacturer, and further jar tests were performed to optimize the coagulant dose. The low dose of ACH not only successfully removed color but also stabilized the membrane filtration. Figure 5.2 and Figure 5.3 present the TMP data of Ashland membrane systems versus time in the same season of 2003 and 2004, respectively. During the membrane operation without ACH addition, membrane fouling became rapid and the TMP increased dramatically after the TMP reached 50% of maximum pressure. However, when ACH was added to the system, the TMP increased steadily an more slowly during the filtration run compared to when ACH was not used.

The fact that the trend repeated for two consecutive years further confirmed the conclusion that low dose of ACH could stabilize membrane performance and slow down the TMP increase. Appendix B includes more details of the Ashland Water Treatment Plant case study.

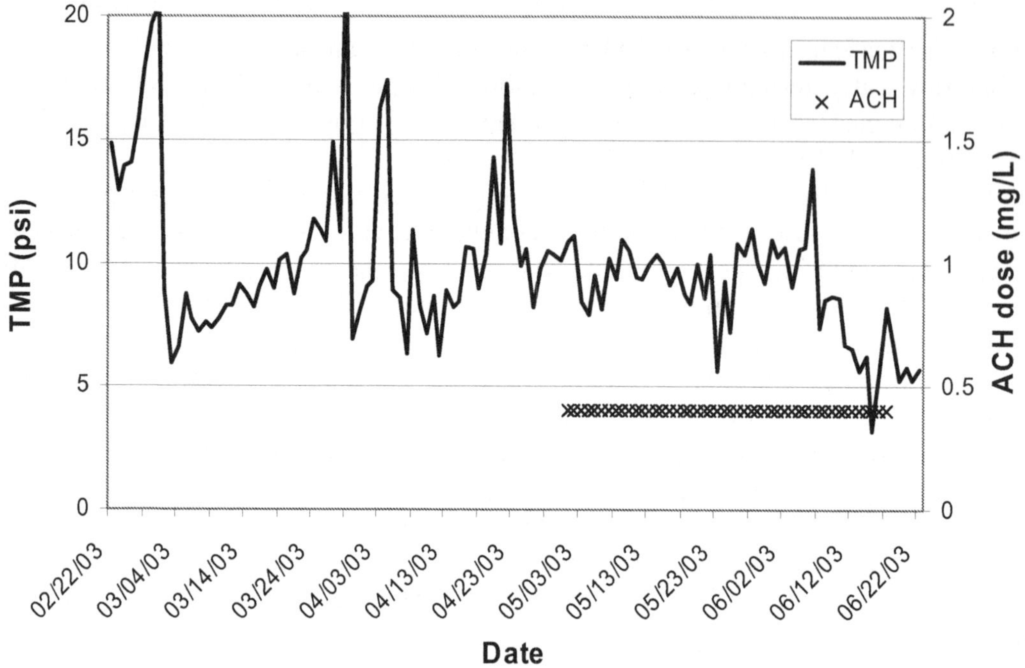

Figure 5.2 Ashland membrane TMP data and coagulant doses in 2003

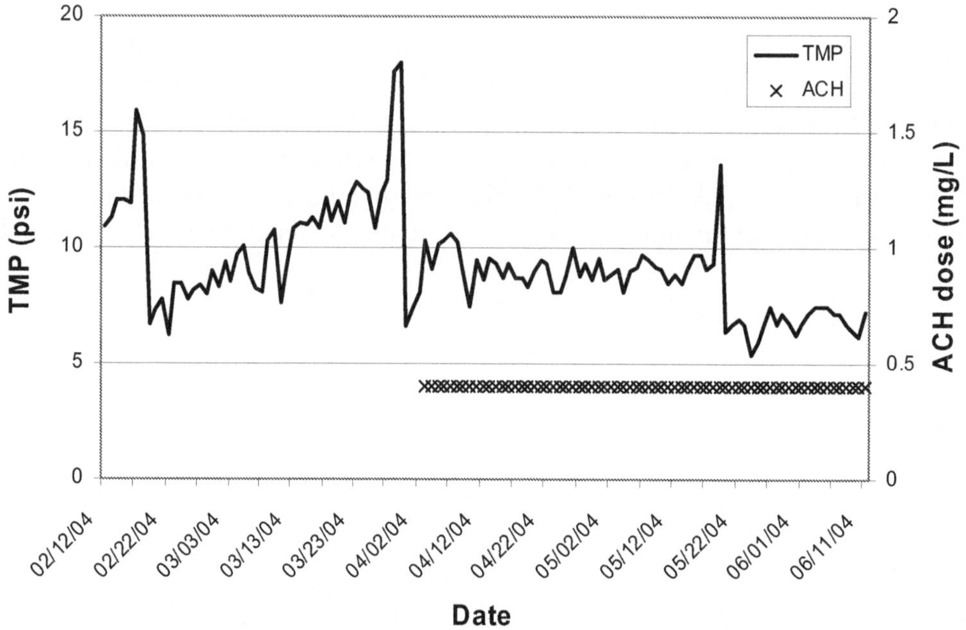

Figure 5.3 Ashland membrane TMP data and coagulant doses in 2004

Effect of Coagulant Type

Both bench- and pilot-scale studies demonstrated that ACH performed better than alum when used as coagulant for clarification. At the same metal mole dose, the specific flux decline rate of a membrane fed with ACH coagulated water was slower than of alum coagulated water for PVdF and PS membrane. The Meredith Water Treatment Plant (Meredith) also provided data further supporting that ACH performed better than alum for membrane pretreatment.

Meredith is located in Victoria, Australia, which is a 0.6 mgd (2.5 ML/day) microfiltration plant with polypropylene membrane consisting of 72 membrane modules.

ACH and acid are dosed directly into the inlet main and passed through an in line mixer, which provides two to three minutes of contact time before the membrane system. Chloramine is used as disinfectant after membrane for water distribution.

The treatment process, including direct coagulation and microfiltration, successfully removes turbidity from 1-14 NTU to less than 0.1 NTU, color from 5-55 TCU to 5 TCU, iron from 2.5 mg/L to less than 0.3 mg/L, and manganese from 1 mg/L to less than 0.1 mg/L. The membrane system is backwashed every 20 minutes, and the backwash waste is transported by gravity from the backwash tank to one of the two 14m x 24m sludge lagoons. Clear supernatant is drawn from the top of the lagoon and returned to the storage basin when the ACH floc settles. The CIP is conducted approximately every two weeks or when the TMP increases to the upper limit, 150 kPa (22 psi).

Alum was initially used as coagulant at Meredith. However, after a series of testing, ACH was found as a better coagulant. Figure 5.4 shows the membrane TMP versus time with alum and ACH as coagulant. ACH reduced the membrane fouling rate by more than 50% compared to the fouling rate when alum was used. With alum, the membrane TMP increased by 60 kPa (8.7 psi) in five days, while the TMP increased 60 kPa (8.7 psi) in more than 10 days when ACH was used. Appendix B includes a more detailed case study of Meredith.

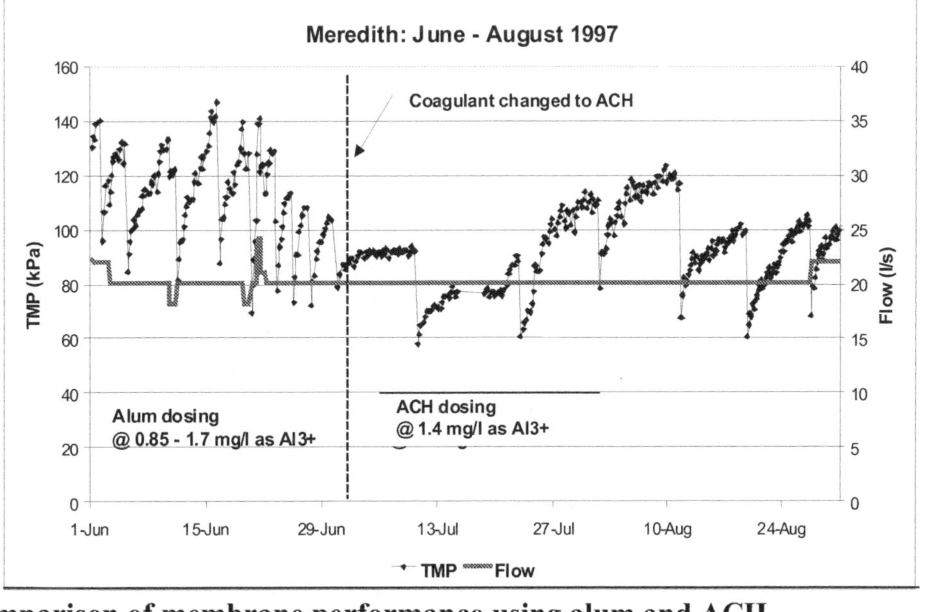

Figure 5.4 Comparison of membrane performance using alum and ACH

Effect of Coagulant and Water Quality Interaction

The Waikato Water Treatment Plant (Waikato) demonstrated the interaction of coagulant and raw water quality. Waikato, located in New Zealand, has a submerged PVdF membrane system with a design capacity of 12 mgd.

Pretreatment to the membrane consists of sludge blanket clarification using alum as coagulant. The plant treats river water and experiences significant spikes in turbidity during storm events. Since the river water has low alkalinity (20 - 45 mg/L as $CaCO_3$), lime and carbon dioxide are added to increase alkalinity to improve coagulation and adjust pH. Granular activated carbon (GAC) is used for final polishing of membrane filtrate. Average pretreatment chemical doses are 2 mg/L lime, 40 mg/L alum, and 15 mg/L carbon dioxide. Pretreatment conditions at this plant have not been significantly changed since plant startup in July 2002.

The objective of pretreatment at this plant is the reduction of particulate and organic foulants. The sludge blanket clarifier has been operated reliably and consistently at removing these foulants with average turbidity decreasing from 18.5 NTU in the raw river water to 1.5 NTU in the clarifier effluent, while DOC is reduced from 4 mg/L on average in the raw water to 2.4 mg/L in the membrane filtrate.

While pretreatment has been successful at removing particulate and organic foulants, it has also been determined to play a role in the inorganic fouling of the membranes. Silica present in the raw water at a high level (averaging 28 mg/L), along with aluminum from the coagulant, combine to form kaolinite. Kaolinite, $[Al_2Si_2O_5(OH)_4]$, was determined through membrane autopsy to be the major foulant.

Although the plant has consistently met its membrane performance objectives, including 30 days between cleanings, it became apparent after approximately one year of operation that kaolinite fouling increased the CIP frequency. To address this, citric acid maintenance cleans were initiated on an as needed basis to extend the run times. On days that citric acid maintenance cleans were performed, daily sodium hypochlorite maintenance cleans (250 mg/L) were not performed.

Figure 5.5 shows the specific flux versus specific volume in 2002 when citric acid maintenance cleans were not employed and 2003 when citric acid maintenance cleans were conducted. These two filtration sets occurred at approximately the same time of year, so the water quality are very similar. The slopes of the best-fit lines represent the specific flux decline rate. The fouling rate (0.0038) for the operation when only daily chlorine maintenance cleans were performed is more than 30 percent higher than the fouling rate (0.0029) when citric acid maintenance cleans were performed on average every 5 days. It is interesting to note that the filtration cycle with citric acid maintenance cleans contained a number of points with specific flux significantly higher than the average value. Often these data points were produced when the hypochlorite maintenance cleans were performed immediately after the day of a citric acid maintenance clean. The hypochlorite maintenance cleans were particularly effective at restoring specific flux at this case.

Even though alum contributes to the inorganic fouling, it was discovered that maintaining alum as the coagulant and operating the system with more frequent citric acid maintenance cleans would be more economical than changing coagulant. Furthermore, alum has successfully removed organic and minimized trihalomethane (THM) formation.

The bench-scale study of Waikato and Clackamas River waters with PVdF membrane (second and third graphs of Figure 3.20) showed that the silica in the raw water could interfere with aluminum-based coagulant to cause inorganic fouling.

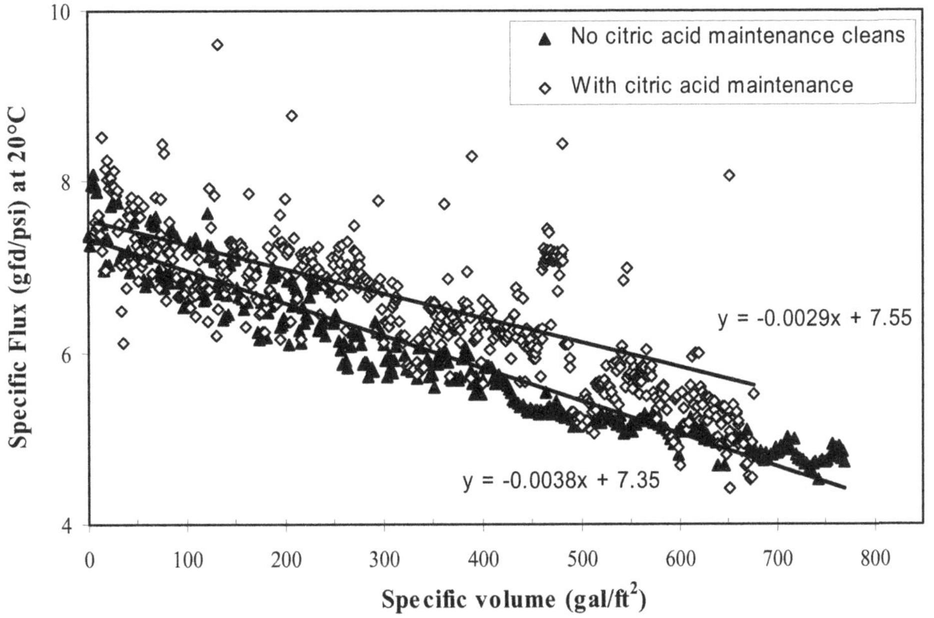

Figure 5.5 Membrane specific flux versus specific volume

CHAPTER 6
COST ANALYSIS

Pretreatment may affect membrane system operation in a number of ways, such as allowing sustained operation with higher flux, lower pumping pressure, less frequent backwashing, reduced cleaning requirements, or increased membrane life. Some design and operational decisions may primarily affect capital costs; others may primarily affect operational and maintenance (O&M) costs. For instance, a lower pumping pressure can reduce operational costs because of lower energy requirements, whereas reduced backwashing and cleaning requirements can reduce operational costs through lower chemical requirements and less labor involved with backwashing and cleaning.

Capital costs can be reduced if pretreatment allows a membrane system to be operated at a higher flux, because the system can be constructed with fewer membrane modules, leading to a commensurate reduction in piping, fittings, valves, and other appurtenances. A previous AwwaRF study (Howe and Clark 2002a) developed a model for assessing the incremental costs of coagulation on membrane filtration, with or without clarification. The model allows the added costs associated with pretreatment to be compared to the cost savings resulting from operation at a higher flux. The model was based on a model that was originally developed by Pickering and Weisner (1993), and was updated based on more recent cost information on the cost of membrane filtration plants.

MEMBRANE COST MODEL

Pickering and Wiesner (1993) developed a membrane facility cost model that allows the cost of a membrane plant to be predicted based on a small number of inputs. A significant feature of the model is that it requires the permeate flux as an input variable, which allows changes in membrane facility cost to be calculated as flux change. To develop the model, Pickering and Wiesner examined cost data from seven NF and UF plants ranging from 0.1 mgd to 6 mgd (380 to 23,000 m^3/day). Costs were divided into several major components, including buildings, chemical feed systems, control and instrumentation, site work, storage, residual disposal, process piping, yard piping, site electric, pretreatment, cleaning, and booster pumps. These costs were used to develop a model that calculates membrane and nonmembrane capital costs separately. Membrane costs include only the initial cost of the membrane modules. Support equipment, such as pumps, valves, skids, and monitoring equipment, are included in the nonmembrane costs. The model calculates the number of modules from the design flow, membrane flux, recovery, and membrane area per module. Membrane and nonmembrane costs are both a function of the number of modules. Membrane costs are directly proportional to the number of modules, but the model predicts a small economy of scale for nonmembrane costs. Costs in the Pickering model were adjusted to a cost basis in 1990 dollars. Membrane facility costs calculated from the Pickering model (using values from the original article) for plants ranging from 0.1 to 100 mgd (380 to 380,000 m^3/day) are shown in Figure 6.1.

Following the publication of the Pickering model, several other surveys of membrane costs have been published, including Adham et al. (1996), Elarde and Bergman (2001), and Adham et al. (2005). Adham et al. (1996) surveyed total capital costs for 28 facilities (24 MF plants and 4 UF plants) and developed a regression equation to fit the data. This regression curve, which has a cost basis in 1993 to 1995 dollars, is also shown in Figure 6.1. The costs in this 1996 data set are significantly lower than the Pickering model because this survey included mostly MF facilities, which may have a lower cost than the NF plants used in developing the Pickering model, and because the 1996 Adham data set excluded land acquisition, site work, source water development, administrative, engineering, and legal costs due to the fact that they can be highly site-specific.

Elarde and Bergman (2001) surveyed 12 MF and UF plants ranging from 0.5 to 25 mgd (1,900 to 95,000 m^3/day) and developed a regression equation to fit the data. Contrary to the surveys conducted by Pickering and Adham, this survey included mostly larger plants, with only one plant less than 1 mgd (3,800 m^3/day). Finally, Adham et al. (2005) surveyed 44 MF and UF plants ranging from 1 to 70 mgd (3,800 to 265,000 m^3/day) and developed a regression equation to fit the total membrane plant cost data, with a cost basis of 2003. These curves are also shown in Figure 6.1, and the regression equations from each model are shown in Table 6.1

Typically, facility costs from different years are adjusted to a common base by multiplying the costs by a ratio of indices for the original and base years to account for inflation. This approach cannot be used for membrane facilities. Membrane treatment costs have declined over the past 15 years because of advances in technology and increased competition, contrary to the general rate of inflation in the construction industry. In membrane treatment, prices from the most recent facilities reflect costs most accurately.

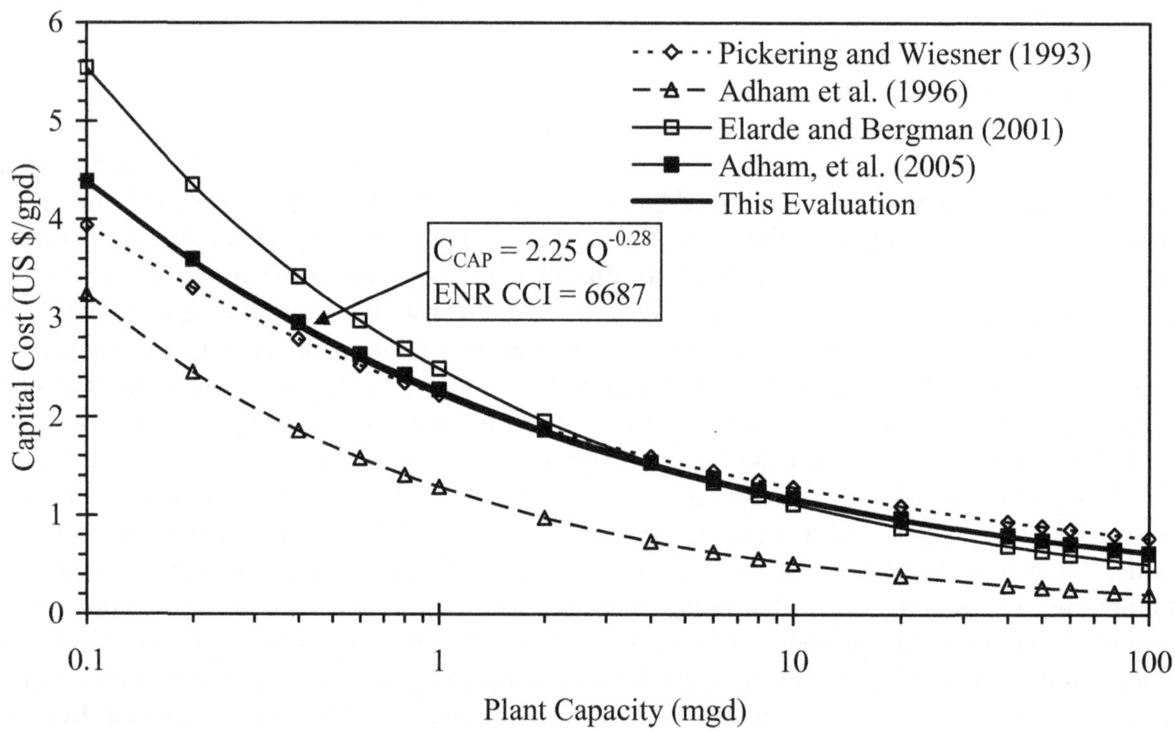

Figure 6.1 Capital cost of membrane facilities

Table 6.1
Cost curves for membrane filtration plants

Data source	Number of plants surveyed	Capital cost regression equation*	Eqn. No.
Pickering and Wiesner (1993)	7	$C_{CAP} = 2.24\ Q^{-0.24}$	(6.1)
Adham et al. (1996)	28	$C_{CAP} = 1.29\ Q^{-0.40}$	(6.2)
Elarde and Bergman (2001)	12	$C_{CAP} = 2.49\ Q^{-0.35}$	(6.3)
Adham et al. (2005)	44	$C_{CAP} = 2.28\ Q^{-0.28}$	(6.4)
This evaluation	—	$C_{CAP} = 2.25\ Q^{-0.28}$	(6.5)

* C_{CAP} = capital cost in \$/gpd.
Q = plant capacity in mgd.

To compare membrane costs with and without pretreatment, a modification of the Pickering model was used. The basic form and the equations of the Pickering model were used, but the values of various parameters were modified so that a regression analysis of data from the Pickering model would match the regression equation from the 2005 Adham data set. The Pickering model was used because it allows flux to be considered when calculating cost, and the 2005 Adham regression equation was used because it is the most extensive and most recent. The equations and parameters in the modified model are presented in Table 6.2. Membrane module size was based on an intermediate-sized model, based on information from manufacturers. The membrane model cost was based on a cost of about \$3.70/ft^2, which was the approximate cost provided by the manufacturers of the pilot plants in this study.

Using the modified Pickering model, membrane facility costs were developed for a range of capacities from 1 to 100 mgd (3,800 to 380,000 m^3/d), and a range of permeate fluxes from 25 to 150 gpd/ft^2 (42.5 to 255 L/m^2·h). The total facility cost using a flux of 50 gpd/ft^2 (85 L/m^2·h) is shown in Figure 6.1 and is described by the regression equation in Equation 6.5. A multivariate regression analysis considering both permeate flux and capacity was used to develop a regression equation that can be used to evaluate the effect of flux at any value of capacity:

$$C_{CAP} = 37.74\ Q^{-0.279}\ J^{-0.723} \qquad (6.11)$$

where: C_{CAP} = capital cost in \$/gpd.
Q = plant capacity in mgd.
J = design permeate flux in gfd.

The relationship between capacity, flux, and capital cost described in Equation 6.11 is shown in Figure 6.2.

Table 6.2
Equations and parameters used in modified Pickering model

Equation	Eqn. No.
$A_M = \frac{Q}{JR}$	(6.6)
$n_{mod} = \frac{A_M}{A_{mod}}$	(6.7)
$C_P = 162000(n_{mod})^{0.70}$	(6.8)
$C_M = C_{mod} n_{mod}$	(6.9)
$C_{CAP} = \frac{C_P + C_M}{Q}$	(6.10)

		Parameter	Value used in model
Q	=	plant flowrate	varied from 1 to 100 mgd (3,785 to 378,500 m^3/d)
J	=	design flux	50 gal/ft^2·d (85 L/m^2·h) used in Figure 6.1, range of 25 to 150 gal/ft^2·d (42.5 to 255 L/m^2·h) used in developing regression
R	=	recovery	92 percent
A_M	=	total membrane area	calculated
A_{mod}	=	membrane area in one module	538 ft^2 (50 m^2)
n_{mod}	=	number of modules	calculated
C_P	=	non-membrane costs	calculated
C_{mod}	=	cost of one membrane module	$2000
C_M	=	membrane costs	calculated
C_{CAP}	=	total capital costs	calculated

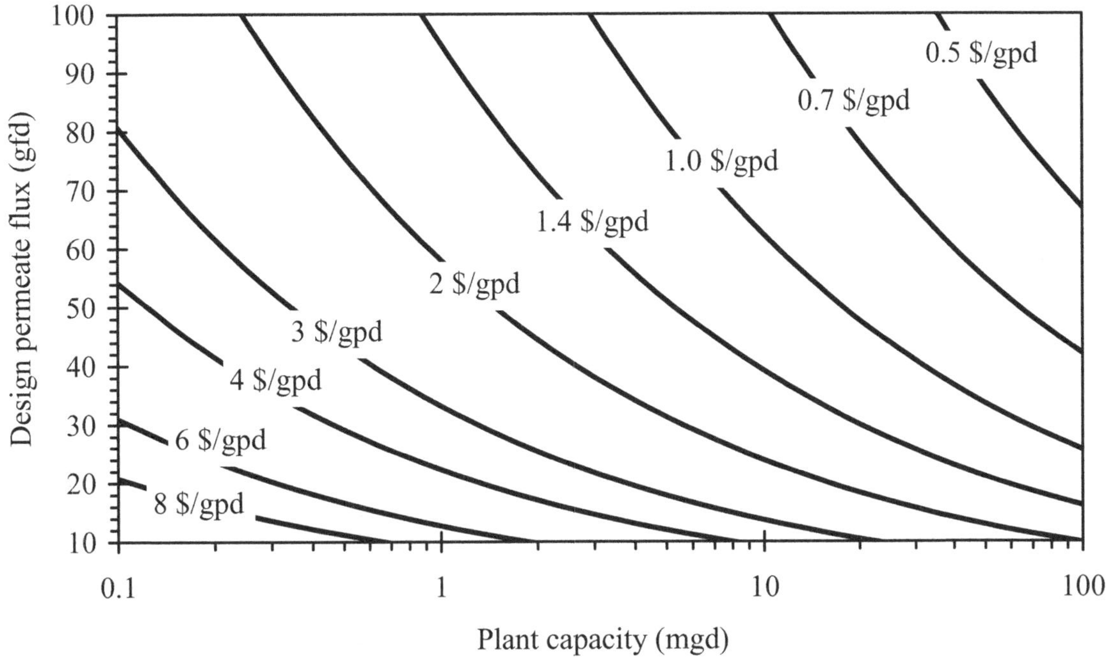

Figure 6.2 Capital cost of membrane facilities based on regression of flux and capacity data (ENR CCI 6687).

PRETREATMENT COSTS

Pretreatment costs were developed for two pretreatment strategies: (1) coagulation/ flocculation without sedimentation, and (2) coagulation/flocculation with sedimentation. Pretreatment costs were generated using cost curves developed by United States Environmental Protection Agency (USEPA) (Gumerman et al. 1979). The original USEPA cost curves were created using unit takeoffs from actual and conceptual designs, cost data from actual plant construction, data from manufacturers, and published data. They were published as a function of a single suitable parameter, such as feed rate (lbs/hour) for chemical feed equipment or basin volume (ft^3) for flocculation basins. Tabular data from USEPA were plotted, a regression equation was determined, and costs were calculated for a range of plant capacities. Overall costs such as land, site work, engineering, contractor overhead and profit, interest during construction, and legal, fiscal, and administrative costs, which are not included in the USEPA costs, were calculated as a percentage of construction cost. The USEPA costs (based on October 1978 dollars) were escalated to mid-2003 costs using Engineering News Record (ENR) Construction Cost Index (CCI) values. The cost was adjusted to 2003 CCI value in order to be compatible with the data set of Adham (2005). The CCI value used was 6687.

The costs for direct filtration pretreatment include chemical storage and feed facilities, rapid mix basins, flocculation basins, sludge lagoons, and overall costs. For processes where cost depend on water quality (such as the sludge lagoons), a turbidity of 20 NTU and a temperature of 15°C was used. Chemical storage and feed facilities were based on use of liquid alum as the coagulant with a dose of 30 mg/L. The costs include fiberglass reinforced polyester (FRP) tanks with a 15-day storage supply, dual-head metering pumps, stainless steel feed piping, and all miscellaneous valves and fittings. Rapid mix costs were based on rectangular concrete basins containing vertical-shaft variable-speed turbine mixers with a detention time of 1 minute and a mixing energy value (G value) of 300 s^{-1}. Flocculation basin costs are based on variable-speed horizontal-paddle flocculators with a G value of 80 s^{-1} in 12-foot-deep (3.7-m) rectangular concrete basins with a detention time of 30 minutes. Residual handling costs are based on a dry solids loading rate of 8 lb/ft^3 (130 kg/m^3), the use of sludge lagoons with a 4-foot (1.2-m) liquid sludge depth and 4-month fill duration, and the assumption that sufficient land is available. Residual handling costs are highly variable, and the use of other sludge disposal technologies may significantly influence the costs of pretreatment. After development of costs using the USEPA model and these values, a regression equation of coagulation/flocculation pretreatment without sedimentation was developed:

$$C_{CF} = 0.283\ Q^{-0.516} + 0.028 \qquad (6.12)$$

where C_{CF} is the cost of pretreatment facilities for coagulation/flocculation without clarification.

The costs for conventional pretreatment are based on the use of upflow solids-contact clarifiers, which contain the rapid mix, flocculation, and sedimentation facilities. The clarifiers were assumed to be 16-foot-deep (4.9-m) circular concrete basins with a mixing energy (G value) of 70 s^{-1}, an upflow velocity of 1 gpm/ft^2 (2.4 m/h), and a detention time of 2 hours. Costs for chemical facilities and sludge disposal were the same as for the direct filtration pretreatment costs. Based on these costs, a regression equation of coagulation/flocculation pretreatment with clarification was developed:

$$C_{CLAR} = 0.695\ Q^{-0.519} + 0.038 \qquad (6.13)$$

where C_{CLAR} is the cost of pretreatment facilities for coagulation/flocculation with clarification.

COST COMPARISON BASED ON PILOT RESULTS

The pilot study using Chattahoochee River water at the AFCWTP demonstrated that appropriate pretreatment can produce an improvement in the net specific flux. In this study, the pilot was operated at 23.3 gal/ft^2·d (39.6 L/m^2·h), and an improvement in net specific flux resulted from a lower operating pressure. The full-scale AFCWTP has a capacity of 96 mgd (363,000 m^3/d). If a membrane facility were build with this capacity and flux, Equation 6.11 predicts a total construction cost of $104 million dollars. If coagulation/flocculation without and with sedimentation were added to the plant without changing flux, the total plant cost would increase to $109 and $114 million dollars, respectively. These costs are detailed in Table 6.3. Also shown in Table 6.3 is the total plant cost with pretreatment, if a modest increase in flux is designed into the plant. These costs are calculated by increasing the flux until the total cost of the plant, including the membrane facility calculated with Equation 6.11 and the pretreatment facilities calculated with either Equation 6.12 or 6.13, is equal to the cost predicted without pretreatment. The cost of the plant is identical to the cost without pretreatment if a 7 percent increase in flux accompanies coagulation/flocculation pretreatment, or a 14 percent increase in flux accompanies clarification.

The improvements in net specific flux observed in the pilot plant were more significant than what is necessary to pay for the pretreatment. After filtration of a volume of about 2000 L/m^2, the net specific flux of raw Chattahoochee River water was about 3.62 gfd/psi (89 $L/m^2 \cdot h \cdot bar$). After filtration of the same volume of flocculated water, the net specific flux was 5.05 gfd/psi (124 $L/m^2 \cdot h \cdot bar$), an improvement of about 40 percent. If the permeate flux were increased by 40 percent (to 32.6 gfd, or 55.4 $L/m^2 \cdot h$), the cost of the plant could be reduced to $87 million, as shown in Table 6.3. Furthermore, the net specific flux after filtration of 2000 L/m^2 of clarified water was 5.75 gfd/psi (142 $L/m^2 \cdot h \cdot bar$), an improvement of about 59 percent over the raw water net specific flux. A 59 percent increase of permeate flux to 37 gfd (62.8 $L/m^2 \cdot h$) would decrease the cost of the facility to $84 million. Based on this analysis, clarification is the most cost effective pretreatment strategy for this water and size of plant. It should be noted that this analysis is based on a low initial flux (23.3 gfd, 39.6 $L/m^2 \cdot h$) and large plant (96 mgd, 363,000 m^3/d), which results in high costs because the regression equation is based on a base flux of 50 gfd (85 $L/m^2 \cdot h$). For a large plant operating at a low flux, the cost of pretreatment can be paid for with relatively small increases in flux, and pretreatment is likely to be beneficial if these small increases in flux can be accomplished.

Table 6.3
Cost of a 96 mgd (363,000 m^3/d) facility treating Chattahoochee River water with various pretreatment strategies*

Membrane plant construction scenario	Flux ($gal/ft^2 \cdot d$) [$L/m^2 \cdot h$]	Flux increase above baseline (percent)	Capital cost ($/gpd)	Capital cost (million $)
Baseline flux				
No pretreatment (baseline)	23.3 [39.6]	--	1.08	104
Coagulation/flocculation	23.3 [39.6]	--	1.13	109
Coagulation/flocculation/sedimentation	23.3 [39.6]	--	1.19	114
Flux increased to match cost of baseline				
Coagulation/flocculation	25.0 [42.4]	7	1.08	104
Coagulation/flocculation/sedimentation	26.6 [45.2]	14	1.08	104
Percent flux increase observed in pilot				
Coagulation/flocculation	32.6 [55.4]	40	0.90	87
Coagulation/flocculation/sedimentation	37.0 [62.8]	59	0.87	84

*ENR CCI 6687

A similar cost analysis was performed based on the results of the Seine River pilot studies (Table 6.4). The Seine River pilot was operated at 65 gfd (110 L/m^2·h) and the corresponding full scale plant has a capacity of 14.5 mgd (54,900 m^3/d). Based on this capacity and flux, Equation 6.11 predicts a total construction cost of $12.7 million dollars. These costs increase to $14.1 million and $15.7 million with the addition of coagulation/flocculation and clarification, respectively, based on costs predicted with Equations 6.12 and 6.13. To offset the cost of pretreatment, an 18 percent increase in flux is needed with coagulation/flocculation, and a 43 percent increase in flux is needed with clarification. These greater increases in flux are harder to accomplish with pretreatment than the modest increases in flux that were needed with the Chattahoochee River. The Seine River pilot plant generated about a 17 percent increase in net specific flux after clarification and about a 50 percent in net specific flux after clarification when a polymer was used with the coagulant. In the case of clarification, the increase in net specific flux observed in the pilot plant is insufficient to offset the added costs of pretreatment, so pretreatment with clarification would not be a cost-effective option. With the addition of a polymer, however, the costs of pretreatment are just offset by a 50 percent increase in permeate flux. In this case, pretreatment may be a practical option, although the cost savings are not significant.

Table 6.4
Cost of a 14.5 mgd (54,900 m^3/d) facility treating Seine River water with various pretreatment strategies*

Membrane plant construction scenario	Flux (gal/ft^2·d) [L/m^2·h]	Flux increase above baseline (percent)	Capital cost ($/gpd)	Capital cost (million $)
Baseline flux				
No pretreatment (baseline)	65 [110]	--	0.87	12.7
Coagulation/flocculation	65 [110]	--	0.97	14.1
Coagulation/flocculation/sedimentation	65 [110]	--	1.08	15.7
Flux increased to match cost of baseline				
Coagulation/flocculation	77 [130]	18	0.87	12.7
Coagulation/flocculation/sedimentation	93 [158]	43	0.87	12.7
Percent flux increase observed in pilot				
Clarification	76 [129]	17	0.98	14.2
Clarification with polymer addition	97.5 [166]	50	0.85	12.4

*ENR CCI 6687

A cost analysis was also performed on the Clackamas River pilot results, assuming a 10 mgd (37,800 m^3/d) plant is to be built nominally operating at 50 gfd (85 L/m^2·h). The construction cost of this plant is predicted to be 11.7 million dollars without pretreatment, 12.9 million dollars with coagulation/flocculation, and 14.2 million dollars with clarification, based on the costs predicted with Equations 6.11 to 6.13. These costs are summarized in Table 6.5. To offset the cost of pretreatment, a 15 percent increase in flux is needed with coagulation/flocculation, and a 39 percent increase in flux is needed with clarification. These increases in flux were not accomplished during the pilot testing of Clackamas River water. Figure 4.13 shows about a 5 percent increase in flux with flocculation and a 10 percent increase in flux with clarification. The capital cost with flocculation and a 5 percent increase in flux is about 12.5 million dollars, or about 0.8 million dollars above the cost of the plant without pretreatment. Similarly, the cost of the plant with clarification and a 10 percent increase in flux is 13.4 million dollars, about 1.7 million dollars above the cost of the plant without pretreatment. Since pretreatment results in a net increase in the capital cost of the plant, even when increases in flux are factored in, pretreatment would not be a cost effective option for the Clackamas River water based on the results of these pilot tests.

Table 6.5
Cost of a 10 mgd (37,800 m^3/d) facility treating Clackamas River water with various pretreatment strategies*

Membrane plant construction scenario	Flux (gal/ft^2·d) [L/m^2·h]	Flux increase above baseline (percent)	Capital cost ($/gpd)	Capital cost (million $)
Baseline flux				
No pretreatment (baseline)	50 [85]	--	1.17	11.7
Coagulation/flocculation	50 [85]	--	1.29	12.9
Coagulation/flocculation/sedimentation	50 [85]	--	1.42	14.2
Flux increased to match cost of baseline				
Coagulation/flocculation	57.7 [98]	15	1.17	11.7
Coagulation/flocculation/sedimentation	69.5 [118]	39	1.17	11.7
Percent flux increase observed in pilot				
Coagulation/flocculation	52.5 [89]	5	1.25	12.5
Coagulation/flocculation/sedimentation	55 [93]	10	1.34	13.4

*ENR CCI 6687

A cost analysis was also performed on the Clackamas River pilot results, assuming a 10 mgd (37,800 m³/d) plant is to be built, nominally operating at 50 gpd/ft² ([illegible] m/h). The construction cost of this plant is predicted to be 11.3 million dollars without pretreatment, 12.8 million dollars with coagulation/flocculation, and [illegible] million dollars with [illegible]. [illegible] on these costs, obtained with Equations [illegible] [illegible]. [illegible]

[illegible]

	[illegible]	[illegible]	Capital cost	Capital cost
Alternative plant construction scenario	[illegible]	[illegible]	[illegible]	[illegible] (Million $)
No pretreatment (baseline)				
	[illegible]			
Coagulation/direct filtration	30		[illegible]	12.[illegible]
	[illegible]			
Coagulation/flocculation/sedimentation	[illegible]		[illegible]	14.[illegible]
	[illegible]			
Clackamas River (as observed in pilot)				
Coagulation/direct filtration	[illegible]	1[illegible]	[illegible]	11.[illegible]
	[illegible]			
Coagulation/flocculation/sedimentation	69.[illegible]	[illegible]	[illegible]	13.[illegible]
	[118]			
Cedar River (as observed in pilot)				
Coagulation/direct filtration	5[illegible]		[illegible]	12.[illegible]
	[89]			
Coagulation/flocculation/sedimentation	55	10	[illegible] 34	13.[illegible]
	[illegible]			

BNR CCI 0687

CHAPTER 7
SUMMARY AND CONCLUSIONS

In this project, the evaluation was conducted to determine the impact of water quality, pretreatment chemicals, chemical doses, membrane material, and membrane configuration on MF/UF membrane performance. The following sections summarize the findings of the bench-, pilot-, and full-scale results.

BENCH-SCALE TESTING

Impact of Water Quality

- No single measured water quality parameter, such as pH, alkalinity, hardness, TOC, UV_{254} absorbance, SUVA, or silica could be correlated alone to the extent of membrane fouling. A combination of several factors, or water quality characteristics that are not adequately described by current parameters, may be responsible for membrane fouling.
- Source waters interacted with different membrane materials in different ways. The Clackamas River water caused minimal fouling of the PVdF membrane and extensive fouling of the PS membrane, whereas the Waikato River caused extensive fouling of both membrane materials.
- FTIR spectra of fouled membranes of different materials tested with different source waters showed different surface chemistry. The FTIR spectrum of Clackamas River water showed several peaks associated with the base PVdF membrane material and the FTIR spectrum of the Waikato River water filtered through the same membrane did not show the similar peaks. This observation suggested that the foulant layer on the membrane with the Clackamas River water was thinner than it was on the membrane with Waikato River water, which was supported by both the flux data and the SEM images.

Impact of Pretreatment

- Clarification removed TOC in the range of 12 to 42 percent in the various source waters.
- Clarification improved the membrane performance of the PS membranes tested with 3 different source waters. The coagulant doses were fairly high for the Medina and Seine Rivers (at 6.87 and 7.09 mg/L as Fe, respectively), but the Clackamas River had a low coagulant dose (0.94 mg/L as Fe) and also had improved membrane performance.
- Clarification did not always improve the membrane performance of the PVdF membranes with the 3 source waters tested. The Chattahoochee River water benefited from clarification, but the Clackamas and Waikato River waters had worse performance after clarification.
- Pretreatment with 2 mg/L of PAC on Seine River water had no impact on membrane performance in the bench scale tests.

- With Seine River water, PS membrane performance improved more with alum than with ferric chloride. With Clackamas River water, however, PS membrane performance was similar with ferric sulfate and alum, and had the best performance with ACH. Specific mechanisms for these differences were not identified in this research, but it is likely that different coagulants remove different foulant species, and that the prevalence of foulant species vary between source waters. Additional research to identify specific interactions between coagulants and foulants is needed. In any case, these results suggest that the selection of the best coagulant to use with membrane filtration is site-specific.
- Chlorine pretreatment had no impact on PVdF membrane performance on Clackamas River water, but caused worse performance with PS membranes on the same source water, at the doses used in this study.

Impact of Fractionation

- Size fractionation provided useful insight into the nature of foulants from various source waters.
- Prefiltration of raw water through 1 μm membranes made almost no change in membrane performance for Medina, Seine, and Chattahoochee River waters (using PS membranes for the Medina and Seine, and PVdF membranes for the Chattahoochee). Membrane performance improved after Clackamas (using PS membranes) and Waikato River waters (using PVDF membranes) were prefiltered through the 1 μm membranes. This suggests that the foulants were all below 1 μm in size in the Medina, Seine, and Chattahoochee Rivers, but that some foulants were larger than 1 μm in the Clackamas and Waikato Rivers.
- Prefiltration with 100 kDa membranes improved the flux through all membranes and all source waters when raw water was used. Fouling was minimal after the prefiltration, and the membranes were able to maintain 80 to 90 percent of the original flux. This suggests that a major fraction of membrane foulants is larger than 100 kDa in all of the waters tested. Additional prefiltration through 10 kDa membranes caused no additional improvement in flux. These results suggest that there is little or no foulant material between 10 and 100 kDa in size, but some foulants are smaller than 10 kDa, since a small amount of flux decline was observed even after the 10 kDa prefiltration.
- Prefiltration through the 1 μm membranes caused worse performance than unfractionated water for PS ultrafiltration membranes filtering clarified Medina and Seine River waters. These results suggest that the floc larger than 1 μm in the coagulated water created a cake layer on the membrane, which provided a removal mechanism for smaller foulants and prevented them from reaching the membrane. Prefiltration removed the floc larger than 1 μm, restricted the formation of the cake layer, allowed the smaller material to reach the membrane, and caused additional fouling. This effect was not seen with Clackamas River water, however.
- Prefiltration through the 1 μm membranes caused a substantial improvement in performance for PVdF membranes filtering clarified water for all waters (Chattahoochee, Clackamas, and Waikato) and coagulants (alum and ACH) tested. The difference in performance between the PS and PVdF membranes suggests

different mechanisms of fouling for these UF and MF membranes. After coagulation and clarification, the PVdF microfiltration membrane appears to be fouled primarily by particles larger than 1 μm, whereas the PS ultrafiltration membrane appears to be fouled primarily by smaller particles. The differences may be due to the differences in pore size.

- After prefiltration through the 100 kDa membranes, all membranes experienced improved performance using coagulated/clarified feed waters with all source waters. The membranes were able to maintain 80 to 90 percent of the original flux, which was similar to their performance during filtration of raw water prefiltered through 100 kDa membranes. These results suggest that coagulation and clarification may have little or no impact on foulants smaller than 100 kDa in size. Coagulation research supports the hypothesis that coagulation preferentially removes larger material.

Hollow Fiber vs. Flat Sheet Membranes

- In general, fouling of flat sheet membranes was more extensive than hollow fiber membranes, when both were exposed to the same feed water. The only exception to this was when Clackamas River water was filtered through PVdF membranes, which may have been due to the minimal fouling of this water with the flat sheet PVdF membranes.
- The flat sheet and hollow fiber membranes generally experienced the same trends when exposed to different pretreatments, such as prefiltration or coagulation and clarification, which suggests that flat sheet membranes are an acceptable surrogate for hollow fiber membranes.

PILOT-SCALE TESTING

Effect of Pretreatment with Coagulants

- Clarification improved membrane performance for pilot-scale tests conducted on Chattahoochee, Seine, and Clackamas River waters.
- Among the different stages of conventional pretreatment (coagulation, flocculation, and clarification), clarified water caused the least amount of fouling, followed by flocculated water, and coagulated water.
- Small changes in the coagulant dose can have a large impact on the membrane performance. Results from clarified Clackamas River water pilot tests revealed that the membrane fouled more rapidly when 0.5 mg/L as Al of ACH was used as coagulant than when 0.2 mg/L as Al of ACH was used. The Clackamas River water with the 0.5 mg/L as Al of ACH fouled the membrane even more than the raw water did, while the same water with 0.2 mg/L as Al of ACH had slightly better performance than the raw water.
- It is recommended that if membrane pretreatment using coagulants is desired, coagulant type and dose be optimized based on site-specific water quality and pilot-scale testing. Optimizing the coagulant dose based on traditional water quality parameters from conventional jar testing (i.e., turbidity, TOC, SUVA, etc.), may not

provide the most accurate optimal dose for membrane pretreatment. The same caution should be taken with addition of polymer.

- Improved membrane performance was observed when raw water was pretreated with ACH compared to alum, at the same metal mole equivalent dose.
- Raw Seine River water coagulated and clarified with the addition of an anionic polymer fouled the membrane less than the water without polymer addition. The polymer type and dose were selected by screening tests.

Effect of Alternative Pretreatment (pre-chlorination and PAC)

- Pre-chlorinating raw Clackamas River water improved membrane performance compared to raw water alone. The improvement was expected to be due to biofouling control that did not affect the bench-scale testing due to the short testing period.
- The addition of 50 mg/L PAC as pretreatment caused more fouling in Clackamas River water compared to raw water. Similar to the conclusion of coagulant dose, the PAC dose should be optimized for site-specific water quality along with other pretreatment processes.

Effect of Membrane Material

- Pilot-scale results of clarified Seine River water demonstrated that CA membranes performed slightly better than PS membranes for this specific water quality using ferric chloride as coagulant.
- Two tests of Chattahoochee River water showed that although both base membrane materials are PVdF, the slight difference of membrane surface chemistry caused different membrane performance.

FULL-SCALE CASE STUDIES

The key findings of the full-scale plant evaluations confirmed pretreatment effects observed in the bench- and pilot-scale studies. The results are summarized below:

- The City of Westminster's Northwest Water Treatment Facility data supported pilot-scale results comparing raw water to clarified water. The specific flux decline rate without pretreatment was more than twice the rate when the membrane system was operated with pretreatment.
- The Ashland Microfiltration Water Treatment Plant data showed that in-line coagulation at a low dose could remove color (seasonal issue) and assist in stabilizing the membrane transmembrane pressure (TMP).
- The Meredith Water Treatment Plant provided data verifying that ACH performed better than alum for membrane pretreatment. At the same metal mole dose, the specific flux decline rate of a membrane fed with ACH coagulated water was slower than that when alum was used as a coagulant.
- The Waikato Water Treatment Plant data demonstrated the interaction of coagulant and raw water quality. At Waikato, high levels of silica in the raw water combined with aluminum from the coagulant to form kaolinite, [$Al_2Si_2O_5(OH)_4$], which was

determined through membrane autopsy as the major foulant. Periodic citric acid maintenance efficiently reduced the membrane inorganic fouling rate.

COST ANALYSIS

The cost analysis was based on a model originally developed by Pickering and Weisner (1993) and updated with more recent cost information from a membrane survey (Adham et al. 2005). The pilot-scale testing results were used in the cost analysis based on the current or future plant capacity. Each case of the cost analysis is summarized below:

- Both flocculation and sedimentation pretreatment are cost effective for a membrane plant with water quality similar to Chattahoochee River. The pilot-scale testing results at AFCWTP using Chattahoochee River water demonstrated that the normalized specific flux increased by 40% when operated with raw water compared to flocculated water. Based on the cost model, a 96 mgd membrane facility will have a capital cost decrease of 16% when flocculation is used as membrane pretreatment. The capital cost will decrease by 19% when sedimentation is used compared to direct filtration with raw water, based on the pilot results which demonstrated that the normalized specific flux of settled water was 59% higher than raw water.
- Sedimentation is not a cost effective pretreatment for a membrane facility with water quality similar to Seine River. The pilot test results obtained at CIRSEE using Seine River water as water source demonstrated a moderate normalized specific flux increase of 17% with settling compared to raw water. A 14.5 mgd plant will cost 11% more with settled water compared to raw water. The flux increase is not high enough to recover the extra capital cost of a settler. However, with addition of polymer at an optimized dose, the capital cost of the membrane plant with settling and polymer addition will be close to the capital cost of a plant without pretreatment.
- Neither flocculation nor sedimentation is a cost effective pretreatment for a membrane plant with water quality similar to Clackamas River. The pilot testing results at NCCWC using Clackamas River water showed slight normalized specific flux increase, 5%, when the membrane system was operated with flocculated water compared to raw water. The cost model predicted 7% increase in capital cost when a 10 mgd plant selects flocculation without sedimentation as pretreatment for membrane compared to direct filtration. A 10% increase in the normalized specific flux was observed during pilot testing when the system was operated with sedimentation compared to raw water (without pretreatment). The capital cost was predicted to be 15% more with the pretreatment process compared to without pretreatment.

CONCLUSIONS

The initial project goal was to develop a unified understanding of the interaction of water quality, membrane materials, pretreatment chemicals, and pretreatment processes. However, as the research was being conducted, the project team discovered that a multitude of variables existed that could impact such interactions, making the original project goal nearly unattainable. Thus, it was decided to focus the bench and pilot-scale testing on selected conditions that were relevant to current applications of MF/UF facilities which require pretreatment. The key findings of the bench- and pilot-scale testing are summarized below (Table 7.1 and Table 7.2) and should be used by utilities, researchers, and the water industry as a preliminary tool to guide their understanding of the impact of pretreatment on MF/UF processes. Future research studies can be built from this tool in order to develop a more comprehensive understanding to optimize pretreatment of MF/UF applications. These findings have also been supplemented with a comprehensive table summarizing selected literature (Appendix A) and case studies from full-scale plants to further assist in understanding the various interactions that impact pretreatment and membrane performance of past and current MF/UF facilities.

In addition to the specific impacts on membrane performance summarized above, the following general conclusions from the bench- and pilot-scale results can be made:

- No single measured water quality parameter, such as pH, alkalinity, hardness, total organic carbon (TOC), UV254 absorbance, specific ultraviolet absorbance (SUVA), or silica could be correlated to the extent of membrane fouling. A combination of several factors, or water quality characteristics that are not adequately described by current parameters, may be responsible for membrane fouling.
- Bench-scale studies typically agreed qualitatively with pilot-scale studies, suggesting that the preliminary membrane selection may be best conducted in the laboratory under more controlled conditions.
- SEM and FTIR analysis provided qualitative results of the membrane performance, which further supported the conclusions of flux decline results.
- Size fractionation provided insight into the nature of foulants from various source waters.
- Future study of fractionation should be used as a tool to further investigate the direct impact of fractionation change by varying pretreatment conditions (e.g., using different coagulants, or different coagulant doses) and the corresponding flux decline.
- The effect of pre-chlorination on membrane performance should be evaluated with long-term testing.
- Coagulant chemical and dose should be optimized based on site-specific water quality and pilot-scale testing. Coagulant dose determined based on traditional water quality parameters from conventional jar testing (i.e., turbidity, TOC, etc.) should be verified for optimal membrane performance.

Table 7.1
Effects of pretreatment on membrane performance at bench-scale

		Medina River	Seine River	Chattahoochee River	Clackamas River	Waikato River
PS	A	– moderate fouling	– high fouling	–	– high fouling	– high fouling
	B	– clarification reduced fouling – alum was better than ferric	– clarification reduced fouling – alum was better than ferric – PAC at 2 mg/L did not improve membrane performance	–	– clarification reduced fouling – ACH was better than alum and ferric – chlorine increased fouling at short term	–
	C	–	–	–	–	–
	D	–	– flat sheet membrane fouled faster than hollow fiber membrane	–	– flat sheet membrane fouled faster than hollow fiber membrane	– flat sheet membrane fouled faster than hollow fiber membrane
	E	–	–	–	–	–
PVdF	A	– moderate fouling	–	– high fouling	– low fouling	– high fouling
	B	– clarification reduced fouling – PACl was better than alum and ferric	–	– clarification reduced fouling	– clarification increased fouling – ACH was better than alum and ferric – chlorine did not improve membrane performance at short term testing	– clarification did not improve membrane performance
	C	–	–	– best membrane performance was observed on clarified water followed by flocculated water, then coagulated water	–	–
	D	–	–	– flat sheet membrane fouled faster than hollow fiber membrane	– flat sheet and hollow fiber membranes fouled at a similar rate	–
	E	–	–	–	– hydrophilic PVdF fouled less than hydrophobic PVdF	–
CA	A	– moderate fouling	–	–	–	–
	B	– clarification reduced fouling – PACl was better than alum and ferric	–	–	–	–
	C	–	–	–	–	–
	D	–	–	–	–	–
	E	–	–	–	–	–

A: impact of water quality B: impact of pretreatment chemicals C: impact of pretreatment processes D: impact of configurations E: impact of membrane materials

Table 7.2
Effects of pretreatment on membrane performance at pilot-scale

		Medina River	Seine River	Chattahoochee River	Clackamas River	Waikato River
PS	A	–	–	–	–	-
	B	–	–	–	–	
	C	–	–	–	–	
	D	–	–	–	–	-
	E	–	– PS fouled slightly faster than CA	–	–	
PVdF	A	–	–	– moderate fouling	– moderate fouling	-
	B	–	–	– clarification reduced fouling	– clarification at low coagulant dose slightly reduced fouling – clarification at higher coagulant dose increased fouling – ACH was better than alum – chlorine reduced fouling at long term testing – PAC at high dose increased fouling	-
	C	–	–	– best membrane performance was observed on clarified water followed by flocculated water, then coagulated water	– best membrane performance was observed on clarified water followed by flocculated water, then coagulated water	-
	D	–	–	–	–	-
	E	–	–	– hydrophilic PVdF fouled less than hydrophobic PVdF	–	-
CA	A	–	– high fouling	–	–	-
	B	–	– clarification reduced fouling	–	–	-
	C	–	– best membrane performance was observed on clarified water followed by coagulated water	–	–	-
	D	–	–	–	–	-
	E	–	– CA fouled slightly less than PS	–	–	-

A: impact of water quality B: impact of pretreatment chemicals C: impact of pretreatment processes D: impact of configurations E: impact of membrane materials

PROJECT RECOMMENDATIONS

The project findings provide guidance to water utilities for integrating MF/UF membrane process into existing or new water treatment plants. It is important to note, however, that results presented in this study may not apply to all cases and situations. Therefore, it is recommended that the overall general conclusions presented herein be used to inform and guide utilities with the most cost-effective and suitable approach (i.e. bench-, or pilot-scale) to address specific concerns and optimize membrane performance.

APPENDIX A LITERATURE SURVEY TABLE

Study	Year	Scale	Source	Membrane	Memb. Details	Study Issues	Pretreatment	Comments
Adham et al.	1991	Bench	Central Illinois groundwater	UF	Aqua.	NOM, 2,4,6-TCP	PAC	Organic removal by PAC-UF system accurately predicted
Adham et al.	1993	Bench, pilot	Seine River, central Illinois groundwater	UF	Aqua.	NOM	PAC	Model predictions successfully verified
Anselme et al.	1999	Full	Seine River	MF, UF	Aqua. (CA-OI-HF)	*Giardia*, *Cryptosporidium*, DBP	Clar., PAC	The process is adaptable to varied flows
Aoustin et al.	2001	Bench	Synthetic, Mooney Dam (Australia)	UF	Millipore (CA-$PP_{support}$-FS)	NOM fouling	$CaCl_2$	Fouling of UF membranes due to different organic matter solute solute interaction understood.
Atencio et al.	1999	Pilot	Utah lake, Jordan River	MF, UF	Memcor MF (PP-P-OI-HF), Aqua.(CE-P-IO-HF)	Turbidity, TOC	None	MF system effectively treated high turbidity with 2 week cleaning frequency
Bergman et al.	2001	Pilot	F. Wayne Hill Water Resource Center (Ga.)	MF, UF	Memcor (PP-S), Koch UF (PS), Zenon (ZEN-S-OI-HF)	Water reclamation	Lime, GAC	UF and MF are feasible for tertiary treatment
Best et al.	2001	Pilot	SW in Canada, Montana, and Central USA	MF, UF	Zenon UF (ZEN-S-OI-HF) LP	Viruses, TOC	Alum, PAC	Lower turbidity, optimized coagulation for DBP removal
Bian et al.	1999	Bench	River water	UF	CA	fouling	PACl	Low doses improved fouling
Bourke et al.	2001	Pilot	Wanneroo (GW) Australia, Atlantic USA	MF, UF	UF, MF (NA)	NOM, Turbidity	MIEX	Fouling reduced, more feasible than RO and NF
Braghetta et al.	1999	Pilot	Ohio River	MF	Memcor (PP-OI-HF-DE)	TOC, fouling	GAC, PACl	In-line coagulation key to remove TOC

Study	Year	Scale	Source	Membrane	Memb. Details	Study Issues	Pretreatment	Comments
Braghetta et al.	2001	Pilot	Medina River (Texas)	MF, UF	Aqua. (XF), Memcor (XF), Koch (DE)	TOC, turbidity, HAA	FER, PAC, Clar.	Less fouling with pretreatment in UF system
Brown et al.	2004	Pilot	Missouri River	UF	Hydra., Polymem, Zenon	Existing system retrofit	PAC	PAC improved flux
Burns et al.	2005	Pilot	Missouri River	UF	UF (S)	Fouling due to solids and polymer carryover	PAC, CP	Carryover polymer did not effect fouling, direct addition made fouling worse
Carroll et al.	2000	Bench	Moorabool River (Australia)	MF	MF (PP-HF-OI)	NOM	Alum, PAC	PAC no reduction in fouling prefiltration coagulation fouling improved
Chang et al.	2001	Pilot	Yellowstone River (Mont.), Tucson (Ariz.)	MF, UF	Zenon UF (ZEN-S-HF-OI), UF (IO-PS), MF (OI-PVdF-hydl), MF (PP-OI-S-HF)	As	FER	Reduction in turbidity arsenic observed though catastrophic result for pressure membrane
Chiu et al.	2005	Bench, pilot, full	Chattahoochee, Medina, Clackamas, Seine, and Waikato Rivers	MF, UF	UF (PS), MF (PVdF GVWP)	Fouling	Alum, FER, ACH	Some cases fouling worse with coagulation others it improved
Chiu et al.	2004a	Bench, pilot, full	Chattahoochee, Seine Rivers	MF, UF	Memcor, Zenon	Fouling	Alum, FER	Coagulation improved flux in bench scale
Chiu et al.	2004 b	Bench, pilot, full	Chattahoochee, Seine Rivers	MF, UF	Memcor, Zenon	Fouling	Alum, FER	Coagulation improved flux in bench scale
Clark et al.	2005	Pilot	Muliple sources	MF, UF	Submerged	Fouling	Alum, PACl, FER, FES	Piloting necessary for implementing pretreatment with membranes, details of piloting vary depending on objectives.

Study	Year	Scale	Source	Membrane	Memb. Details	Study Issues	Pretreatment	Comments
Cleveland and Crozes	1999	Pilot	Willapa River (Wash.)	MF, UF	Memcor (PP-OI-HF-DE), Aqua. (CE-IO-HF-XF)	*Giardia*, *Cryptosporidium*, THM	None	Membranes provide an absolute barrier against any water contaminant
Colas et al.	2002	Pilot	Synthetic	UF	Aqua. (Ultrazur 450)	New membrane development	None	Described the development of a new membrane module
Crozes and Seacord	1999	Pilot	Missouri River (St Louis)	UF	Koch (PES-IO-HF-XF), Aqua. (CA-IO-HF-XF)	TOC, Turbidity, SOC, *Giardia*, *Cryptosporidium*	PAC, Ozone	Lowered turbidity and SOC with minimal fouling
Crozes et al.	1999	Pilot	Lake Winnebago (Wis.)	MF, UF	Aqua.(CA-P-IO-HF-hydl), Koch (PS-P-IO-HF), Memcor (PP-P-OI-HF), Zenon (ZEN-S-OI-HF)	*Cryptosporidium*, turbidity	Lime	Feasible treatment for rem. DBP, microrg.
Curl et al.	2003	Pilot	Lake Decatur (Ill.)	UF	UF (HF-PVdF)	NOM, pathogens	Alum	Coagulation increased virus removal and decreased fouling
Do-Quang et al.	2004	Pilot	Seine River	UF	Aqua. (HF-CA-PS)	NOM	Alum, FER	Flux improved after coagulation and citric cleaning
Durand-Bourlier et al.	1999	Pilot	Seine River	UF	Aqua. (CA-IO-HF-XF)&(CA-IO-HF-DE)	Turbidity, TOC, algae	None	Crossflow can increase flux by 20%
Durand-Bourlier et al.	2001	Pilot, full	Medina River (Texas)	MF, UF	Aqua. (CA-HF-IO-P-hydo), Memcor (PP-OI-HF-hydo), Koch (PES-IO-HF-hydo)	TOC, HAA, viruses, turbidity	FER, PAC	UF (PAC) removes *Giardia*, turbidity, is polishing trt
Elzo et al.	1998	Bench	Synthetic	MF	membralox SCT company	Silica	pH adjustment	Flux dependent upon pH for silica deposition
Fan et al.	2001	Bench	Maroondah Aqueduct, Moorabool River	MF	GVWP and GVHP millipore	NOM, Ca, Na	N/A	Higher aromatic component and high molecular weight fraction caused higher fouling

Study	Year	Scale	Source	Membrane	Memb. Details	Study Issues	Pretreatment	Comments
Fu and Dempsey	1997	Bench	Dismal Swamp	UF	PS	Fulvic acids	Alum	Low doses made fouling worse
Gallagher et al.	2005	Pilot	Creek in north eastern USA	MF	Memcor CMF-S	Mn	ACH, FER	Manganese was removed clarification improved flux
Galloway et al.	2003	Full	Seawater	UF	UF	UF pretreatment for seawater RO	FER, NaOCl	Improved RO process
Garcia-Aleman et al.	2005	Pilot	Lake Ontario	MF, UF	Zenon S, Memcor CMF, Norit	Ions and microorganisms	Ozone, BACC	BACC and ozone work as a barrier for microorganisms
Habarou et al.	2001	Bench	Ribbou Reservoir, Loire River, Blavet, Suwannee River	UF	RC, PS	NOM fractionation according to size and functional groups	N/A	Colloidal fractionation results in more significant fouling
Hackman et al.	1999	Pilot	Lake Decatur (Ill.)	MF, UF	UF (PA, CA)	Nitrates, atrazine	PAC, Alum	UF not suitable prtrt. For removal nitrazine, for RO
Hagmeyer et al.	2003	Pilot	Reservoir water	UF	UF	Particles	$KMnO_4$, Alum	Detention time between coag and UF improved performance
Hall et al.	2003	Full	UK surface waters	MF, UF	Memcor (MF), Kalsep, Norit X-flow filtration	*Cryptosporidium*	Varies	Survey of plants in UK indicate high confidence in Cryptosporidium removal
Hibbs and Kendrick	2003	Pilot	Oak Creek Reservoir, Lake Trammel, Lake Sweetwater	UF	Zenon (ZEN-S-HF-OI)	NOM, TOC, DBPs, turbidity	Alum, ACH, EC	TOC reduced at 30 mg/L coag
Hillis et al.	2003	Full	Ennerdale, Crummock Lakes	MF	Memcor (CMF)	*Cryptosporidium*	PACl	Cost intensive, plant integrity maintained, no residual
Howe and Clark	2001	Bench	Lake Decatur, Salt River, Ohio River	MF, UF	CA, PES, PP	NOM, fouling	Alum, FER, PACl	Coagulation at doses for enhanced coagulation improved membrane performance

Study	Year	Scale	Source	Membrane	Memb. Details	Study Issues	Pretreatment	Comments
Howe and Clark	2002a	Bench	Lake Decatur, Medina River, Ohio River, Beaver lake	MF, UF	CA, PES, PP	NOM, fouling	prefiltration	Fouling was site specific in membranes some fractionation fouled PS membranes did not foul PP membranes
Huey et al.	1999	Pilot	Salt River Project	MF, UF	Memcor (PP-OI-HF-DE), Aqua. (CE-IO-HF-XF)	DBP, *Cryptosporidium*	PAC	Both the membranes get fouled because of minerals in dissolved form
Humbert et al.	2005	Bench	Villejean WTP	UF	RC	NOM	MIEX, IX	Membrane performed same with both
Jack and Clark	1998	Pilot	Lake Decatur, Illinois	UF	CA	Flux, backwashing	PAC, FER	Ferric dose restored performance after period of extensive fouling
Jacangelo et al.	1999	Pilot	Surface water Northern Virginia	MF, UF	Memcor MF (PP-P-HF), Aqua. UF (CE-P-HF)	Pretreatment for NF	MF and UF pretreatment for NF	Permeate flux greater impact than recovery on mem. fouling
Judd and Hillis	2001	Pilot	Synthetic	MF	Memcor (MF-PES-IO-hydl)	NOM fouling	FER	To find optimum coagulant dose for cake layer to prevent NOM fouling
Kader	2005	Full	Ground water	MF	Pall (PVdF-HF)	As	FER	Arsenic effectively removed
Kelly et al.	2003	Pilot, full.	Medina River, Seine River, Kanawha Canal	UF	Aqua.	NOM	Alum, FER	Coagulation improved flux 50 to 100 percent
Kweon and Lawler	2001	Bench	Lake Austin, Missouri River	UF	XF	Hardness	Lime	Recarbonation more significant than lime softening
Lahoussine-Turcaud et al.	1990a	Pilot	Seine River	UF	XF-CA	Colloids	FER	Polysaccharides important in fouling
Lahoussine-Turcaud et al.	1990 b	Pilot	Seine River	UF	XF-CA	Fouling	FER	Polysaccharides important in fouling

Study	Year	Scale	Source	Membrane	Memb. Details	Study Issues	Pretreatment	Comments
Laîné et al.	1989	Bench	Synthetic lake water	UF	RC, PAN	Fouling	PACl	Coagulation helped hydrophilic membrane more than hydrophobic membrane
Laîné et al.	1990	Bench	Synthetic lake water	UF	RC, PAN	Fouling	PACl, PAC	Little impact of PAC, larger impact of coagulation
Lee et al.	2000	Bench	Surface water	MF	N/A	Filtration mode	Alum	Different fouling observed under charge neutralization and sweep floc conditions.
Lee et al.	2001	Bench	Horse Tooth Reservoir, Orange County ground water	UF	UF NTR 7410 XF	NOM fouling and cleaning	NaOH	SDS not effective for NOM cleaning Hydrophobic NOM cleaned easily than hydrophilic
Lee et al.	2003	Bench	Cazaux River, Marne River (France)	MF, UF	CA, PVdF, RC, PES	NOM	N/A	NOM causes fouling by gel layer formation
Levesque et al.	2003	Pilot	Flat Creek (Ga.)	MF, UF	Zenon (PVdF-HF)	*Giardia*	Alum	To dev. Design criterion for integrated systems UF had > ln rem for disinfec.MF >rem for virus / bacteria
Lewis et al.	2001	Pilot, bench	Lake Ivie (Texas)	MF, UF	Pall (encased MF), Zenon (ZEN-S-OI-HF), Memcor (S-MF)	DOC, TOC	FER	UF and MF pretreatment is feasible for RO desalination
Lin et al.	1999	Bench	Synthetic	UF	PS HF XF Hydo	Fouling	PAC, prefiltration	PAC made fouling worse and humic acids facilitate permeate flux
Lin et al.	2001	Bench	Synthetic	UF	HF single fiber neg charge PS 100kDA mwco	pH, conductivity, carboxylics, phenolics	PAC	Humic acid not removed by membranes, PAC made fouling worse
Liu et al.	2005	Bench, full	Surface water (NE Bakersfield water treatment plant)	MF	N/A	Fouling and cleaning protocol	FER	Membrane fouling index (MFI) developed to compare fouling by various pretreatment strategies

Study	Year	Scale	Source	Membrane	Memb. Details	Study Issues	Pretreatment	Comments
Maartens et al.	1999	Bench	Synthetic	UF	Module 10 capillary polysulfone membranes	NOM fouling of HA, FA	$AlCl_3$, $CaCO_3$	pH adjustment min fouling, metal ion treated feed water caused more fouling
Marwah et al.	2004	Bench	Chattahoochee, Medina, Seine Rivers	UF, MF	Aqua. (PS), Millipore (PVdF) HF and FS	NOM	Alum, FER	Coagulation improved flux
McAleese et al.	1999	Pilot	Colorado River	MF, UF	MF [Memcor (PP-OI-HF-LP), Memcor (PP-OI-HF-HP), Zenon (ZEN-S-HF-OI), Pall (PVdF-OI-LP), Pall (PVdF-OI-HP)] UF [Hydra.(PES-IO), Aqua.(CA-IO)]	DBP, Turbidity, TOC	UF and MF pretreatment for NF	Provided the pH is adjusted to 7 it can provide feasible pretreatment
McClelland et al.	2005	Pilot, full	Perdue treatment (sweetwater)	MF, UF	Memcor, Zenon	ClO_2	ClO_2	Design can easily be upgraded
Miller et al.	2005	Pilot	City of Pflugerville	MF	Memcor, Zenon, Pall	Design criteria	Coag.	20-30 min of contact time needed for coagulation
Mourato et al.	1999	Bench	Lake Ontario (Canada), Huntsville, Croton Lake (N.Y.) Sioux (Fla.)	UF	Zenon (ZEN-S-OI-HF)	Fouling, TOC, color	EC, FER, PAC	Color removal up to 95%, TOC removed 75%, lower chemical sludge produced lower chemical use
Mysore et al.	2003	Bench, pilot	Chattahoochee River	MF	Memcor (MF-PVdF-HF)	NOM, Virus, Bacteria psuedomonas	N/A	NOM caused greater fouling and greater microorganism removal

Study	Year	Scale	Source	Membrane	Memb. Details	Study Issues	Pretreatment	Comments
Neemann et al.	2001	Pilot	Lake Brewer, Cadron Creek (Ark.)	MF, UF	Zenon (ZEN-S-OI-HF)	TOC, HAA, THM, DBP, Fe, Mn, *Giardia*	$KMnO_4$, PAC	Effective removal for coli, *Giardia*, TOC, THM's
Norton et al.	2004	Full	Columbia River	UF	Zenon, Memcor PVdF	Existing system retrofit	Coag., Floc., Clar.	Plant capacity expanded without building new basins
Pappas et al.	2003	Bench, pilot	Guadalupe River	UF	Pall (pressure) and Zenon (ZEN-S-HF-OI)	NOM, TOC, DBP	EC, Alum, FER	Submerged inc results for inline coagulation
Paxman et al.	2001	Pilot	Rockport Reservoir (Utah)	MF, UF	Memcor (PP-OI-P), Koch UF (PES), Zenon (ZEN-S-OI-HF)	TOC, DBP	FER, GAC, PAC	7-15% rem TOC, DBP , 70% Fe, 48% SDS THS
Richards et al.	2001	Pilot	Biscayne Aquifer (Fla.)	UF	PP	Fouling	antiscalant	Pretreatment made fouling worse
Roach and Vickers	2001	Full	Nueces River	MF, UF	Memcor, Pall		Alum	Alum made fouling worse
Robert et al.	1999	Pilot	Maumee River (Ohio)	UF	CA	Flux	FER	Coagulation eliminated fouling
Roquebert et al.	2001	Pilot	Kern River (Calif.)	MF, UF	Memcor (PP-OI-HF), Hydra. (PES-IO-HF), Aqua. (CA-IO-HF)	*Cryptosporidium*, *Giardia*, DBP	PACl, NaOCl, $KMnO_4$	Aquasource highest flux
Sahely at al.	2004	Pilot	Lake St. Clair	UF	Zenon	Existing system retrofit	Alum	Pilot testing that pretreatment was not necessary
Schäfer et al.	2001	Bench	Suwannee River humic and fulvic acid	MF, UF	N/A	Fouling	FER	Ferric chloride significantly increased fouling of UF membranes
Schneider et al.	2001	Bench, pilot	Monroe, Alcovy Rivers (Ga.)	MF	PVdF-HF	Mn, turbidity	Ox., Cl_2, ClO_2, $KMnO_4$	Insoluble Mn removed after oxidation
Shrive et al.	1999	Bench, pilot	Ohio River	MF	Memcor	TOC, turbidity	GAC, PACl	In-line coagulation improved treatment

Study	Year	Scale	Source	Membrane	Memb. Details	Study Issues	Pretreatment	Comments
Sethi et al.	2003	Full	Lakes in Calif., Wis., Texas, France, Netherlands	MF, UF	N/A	Turbidity, particles	N/A	Compared pressure decay methods of integrity testing
Sharp et al.	2005	Pilot	N/A	UF, NF	CQc(GE), NF (Hydra.)	Performance with various pretreatments	Coag.	In-line coagulation versus conventional coagulation, formation of dynamic layer increases flux
Shaw et al.	2003	Pilot	Lake Texoma (Texas)	MF, UF	Ionics (PES/PVP-HF), Pall (PVdF-HF), Zenon (ZEN-S-HF-OI)	Turbidity, TOC, taste, color and pathogen removal	MF, UF pretreatment to RO	Memb. fouling same results, once pump in use >deter best method for turbidity rem
Shorney et al.	2001	Pilot, bench	Salt River Project, Central Ariz. Project	MF, UF	S, P	As	FER, acid	Addition of ferric salts made system unstable
Singh et al.	2003	Pilot	Surface water	UF	Zenon (ZEN-S-HF-OI),	TTHM, HAA, DBP, Fe, Mn	PACl, $KMnO_4$, Alum	System needed optimization as the feed changes
Taylor et al.	2003	Pilot	Lake Kickapoo, Lake Arrowhead (Texas)	MF, UF	Zenon (OI-HF) , Pall (OI-HF), Memcor (OI-HF)	Turbidity, *Giardia*, SO_4 in brackish water	FER, Clar.	Pilot conducted for design criterion for brackish water
Voges et al.	2001	Bench	Sludge sample from Boeing contain Cd, Cr, Cu, Pb, Zn	MF	PP-XF	Cd, Cr, Cu, Pb, Zn	HIOP	Addition of iron oxide removed metal from good quality water, and improved flux
White et al.	2004	Pilot	Lake Michigan	UF	Zenon	existing system retrofit	Coag., Floc., Clar.	Conventionally filtered water produced 20 percent higher flux than settled water
Yuan and Zydney	1999	Bench	Synthetic	MF	PES omega series by Filtron tech corp	NOM fouling	N/A	Humic acid adsorpton caused only 10% reduction in filtrate flux, inc fouling during actual humic acid filtration due to deposition on upper surface

Study	Year	Scale	Source	Membrane	Memb. Details	Study Issues	Pretreatment	Comments
Yuan and Zydney	2000	Bench	Aldrich Humic acid, Suwannee River humic acid	UF	PES	NOM fouling	N/A	Prefiltration reduced fouling, large particles have less effect on UF membranes
Zhang et al.	2003	Bench	Synthetic	UF	HF (IO)	SiO_2, fouling	HIOP, PAC	HIOP form a porous layer on the membrane and helps inc. flux

Abbreviations:

Manufacturer

Koch	Koch Membrane Systems
Aqua. Systems	Aquasource Membrane
Memcor	Memcor, US Filter
Zenon	Zenon Municipal Systems
Pall	Pall Microza
Hydra.	Hydranautics

Membrane Material

CA	Cellulose acetate
PA	Polyamide
PAN	Polyacrylonitrile
PES	Polyethersulfone
PP	Polypropylene
PS	Polysulfone
PVdF	Polyvinylidene fluoride
RC	Regenerated cellulose
ZEN	Zenon proprietary
N/A	Not available

Membrane Type

HF	Hollow fiber
FS	Flat sheet
IO	Inside out
OI	Outside in
S	Submerged
P	Pressure
LP	Low pressure
hydo	Hydrophobic
hydl	Hydrophilic
XF	Cross flow
DE	Dead end

Pretreatment

Alum	Coagulation with alum
Clar.	Clarification
Cl_2	Chlorine
ClO_2	Chlorine dioxide
Coag.	Coagulation with unlisted coagulants
CP	Cationic polymer
EC	Enhanced coagulation
FER	Coagulation with ferric chloride
Floc.	Flocculation
GAC	Granulated activated carbon
$KMnO_4$	Permanganate
Lime	Lime softening
MIEX	A micro-size macro-porous strong base magnetic ion exchange resin developed for reversible removal of charged organic ions)
NaOCl	Hypochlorite
Ox.	Multiple unlisted oxidants
PAC	Powdered activated carbon
PACl	Polyaluminum chloride

APPENDIX B FULL-SCALE CASE STUDIES

CASE I: WESTMINSTER NORTHWEST WATER TREATMENT FACILITY

Plant Description

The Westminster Northwest Water Treatment Facility (Westminster) is a 15 mgd Pall Microza microfiltration (MF) plant with 100% feedwater recovery. There are a total eight skids with PVdF pressurized membrane modules. The plant has a unique setup in which the membrane systems can be operated with or without pretreatment. The pretreatment process can include coagulation, flocculation, or sedimentation using PACl as coagulant. When the membrane system is operated with pretreatment, backwash water is transported to the front of the plant (i.e., the backwash water is recovered by the pretreatment process). During seasons when the water quality is above average, the membrane system is usually operated without any pretreatment. The direct membrane filtration utilizes seven skids for operation and the remaining skid is used as a secondary membrane to treat the backwash water.

When the full treatment train is used, raw water is coagulated, flocculated and clarified before membrane filtration. Potassium permanganate followed by sodium hydroxide, and PACl can be added to the raw water before it is passed to the mixer. Then, sodium hypochlorite can be added to the water before flocculation. Flocculated water is then clarified and passed through a 380 micron prefilter before being fed to MF membrane. Figure B.1 presents a schematic of this plant. The dashed line indicates that the plant is capable of operating either with coagulation only, with coagulation then flocculation, or with coagulation, flocculation and sedimentation.

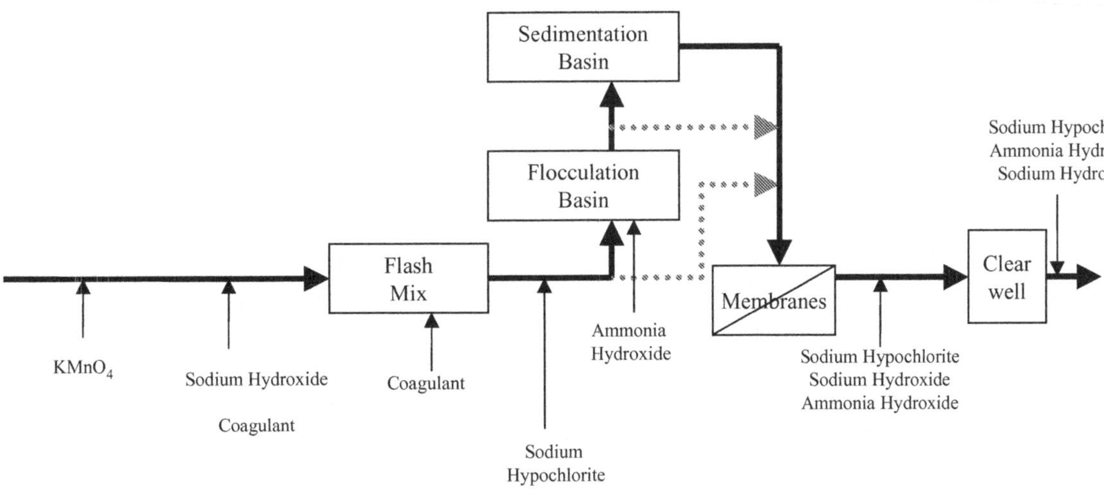

Figure B.1 Westminster process schematic

Water Quality

General water quality of this plant is shown in Table B.1. The raw water has low to moderate alkalinity of about 45 mg/L as $CaCO_3$. As the source is lake water, the raw water turbidity values fluctuate between 1.9 – 16.8 NTU, with an average value of 6.2 NTU. The average TOC concentration is 1.9 mg/L indicating low organics presence. The average TDS concentration in the source water is 175 mg/L.

Table B.1
Westminster water quality

	Units	*Average*	*Range*
Raw Water			
Temperature	°C	12	4 - 15
pH		8.1	7.8 - 8.4
Alkalinity	mg/L as $CaCO_3$		45-55
Turbidity	NTU	6.2	1.9 - 16.8
Color	Standard Color Units	1	0 - 2
TOC	mg/L	1.9	1.8 - 2.0
Iron	mg/L	0.2	0.001 - 0.05
Manganese	mg/L	0.03	0.01 - 1.0
Clarified Water			
Temperature	°C	12	4 - 15
pH		8	7.5 - 8.3
Turbidity	NTU	1.1	0.7 - 3.8
Membrane Filtrate			
Temperature	°C	12	4 - 15
TOC	mg/L	1.8	1.7 - 1.9
Turbidity	NTU	0.012	0.011 - 0.2

Membrane Operation

The Pall microfiltration system consists of 8 skids, of which only 7 are in service at any time. Each skid contains 82 individual filtration modules. The plant uses 2-minute air scrub and reverse flush (AS-RF) enhanced backwash procedure every 20 minutes with approximately 50 mg/L of chlorine dosed in the air scrub water. The reserved skid comes into service when any one of the 7 operational skids is under AS-RF process. The cleaning interval of 20 minutes was used throughout the 4 months of operation irrespective of the pretreatment being used. A steady decline in specific flux was observed throughout the operational period, and chemical cleaning was used at a point when TMP reached 30 psi. Chemical cleaning is performed 8 to 12 times a year, with 18,000 mg/L citric acid with a 60 min soak. Membrane integrity can be indirectly monitored by measuring filtrate turbidity and particles. Direct integrity monitoring using an air pressure hold test is conducted weekly. Approximately 15 fiber breakage incidents occurred within the first 3 months of startup, and an average of 6 fiber breakage incidents occurred per year.

Effects of Pretreatment

When pretreatment is used, PACl is added as the coagulant in the range of 1-12 mg/L with an average value of approximately 5.5 mg/L. Because high manganese concentrations were observed in summer time, potassium permanganate is added seasonally to control manganese levels. Sodium hydroxide is added to increase pH/alkalinity of the raw water and to subsequently improve the flocculation and sedimentation efficiency of the treatment process, if required. After flash mixing with coagulant, water is chlorinated and passed through a serpentine 350,000-gallon flocculation basin employing a horizontal flocculator in each of its 4 chambers. Flocculated water is then clarified by a 350,000-gallon sedimentation basin with stainless steel plate settler technology. Turbidity is reduced from an average of 6.2 NTU in raw water to 1.1 NTU in clarified water. Raw water turbidity and clarified water turbidity are plotted along with PACl dose in Figure B.2. Although the raw water turbidity increased steadily from March to June, the clarified water turbidity was consistent at 1.3 NTU except during the period when pretreatment was bypassed (4/10/04 – 5/5/04), when an average coagulant dose of 5.5 mg/L was used.

Pretreatment was varied during the operational period to evaluate the impact on membrane with and without pretreatment process. Figure B.3 presents the variation in specific flux during the entire operational period. Impact of the pretreatment process is discussed in the following sections by dividing the operational time with or without pretreatment in different phases.

Phase 1: Complete pretreatment including coagulation, flocculation and sedimentation

Between 2/28/04 to 4/9/04, the plant was operated with complete pretreatment including coagulation, flocculation and sedimentation. Raw water turbidity increased steadily from 6 NTU to 10 NTU from February to April and the coagulant dose varied from 6 mg/L to 11 mg/L. With this coagulant dosage, the average clarified water turbidity was maintained at about 1 NTU. The specific flux decreased steadily from 14 gfd/psi to 6 gfd/psi in this phase.

Phase 2: Pretreatment ceased completely

Coagulation, flocculation and sedimentation processes were bypassed between 4/15/04 and 5/5/04 and raw water was fed directly to the membrane after passing through a 380 micron filter. Feed water turbidity increased along with the raw water turbidity (11 NTU to 13 NTU) as no pretreatment was employed. The specific flux decreased significantly from 4 gfd/psi to 1 gfd/psi in 20 days. The rate of specific flux decline was 0.43 gfd/psi/day compared to 0.2 gfd/psi/day when complete pretreatment was utilized in Phase 1.

During this phase, the membranes fouled to the extent that even after resuming the complete pretreatment process on 5/6/04, the specific flux could not be recovered. Hence a CIP with citric acid (pH 2-2.5) was performed on 5/18/04 to restore the permeability of the membrane.

To evaluate the impact of flocculation and sedimentation separately, these units were bypassed during the time period of 4/10/04 to 4/14/04. During this time span, coagulant was added to the system without flocculation and sedimentation. Since the induction mixer was not available during the testing, the coagulation process depended on pipeline turbulence for mechanical flash mixing. Average raw water turbidity was observed at about 11 NTU and with an average coagulant dose of about 4 mg/L. Because there was only 4-day of operation and with only one data point a day, the data was not enough to make solid conclusion.

Phase 3: Second cycle of complete pretreatment after CIP

After a CIP was performed, membrane permeability was restored significantly. During 5/19/04 to 6/30/04, the plant was operated with the complete pretreatment process. Raw water turbidity varied betwen 12 to 14 NTU, and the average clarified water turbidity was 1.4 NTU when an average coagulant dose of 5.5 mg/L was used. The specific flux declined steadily from 11.1 gfd/psi to 3.6 gfd/psi during this phase.

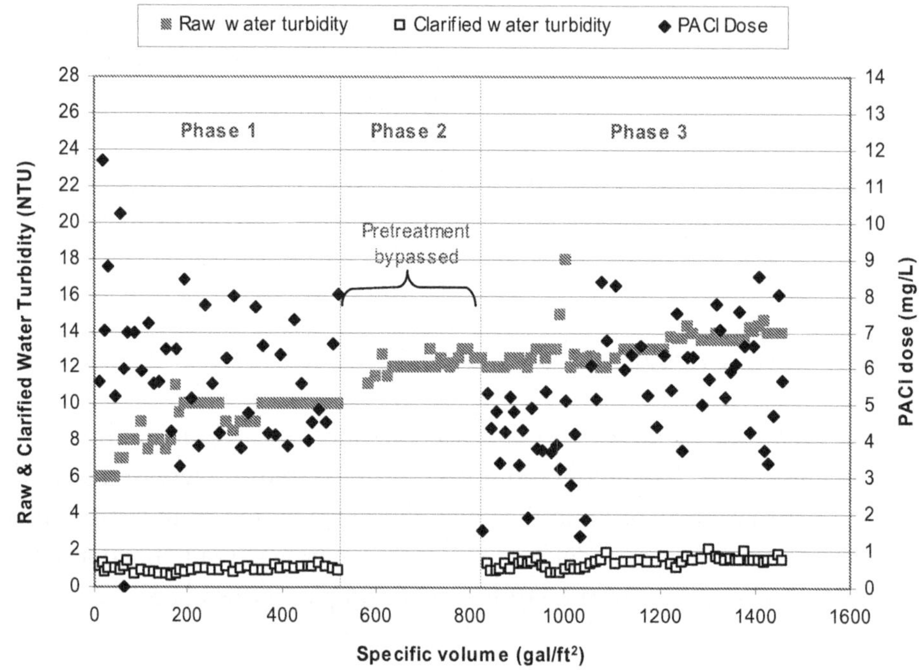

Figure B.2 Turbidity and coagulant dose

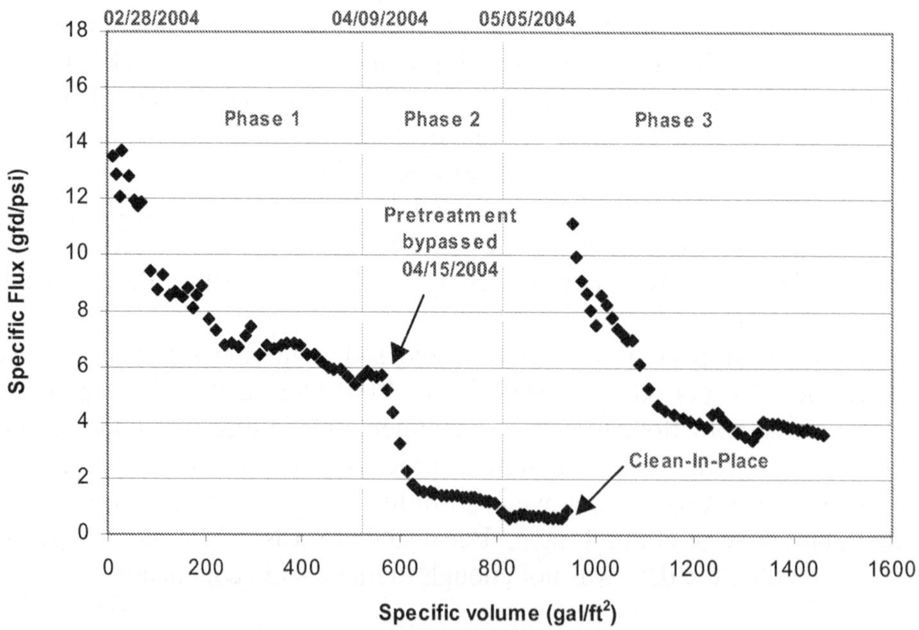

Figure B.3 Evaluating the impact of each pretreatment unit process

Lessons Learned

- The addition of chlorine to the AS-RF water effectively controlled organic-based fouling, reducing the frequency of CIPs required. The source water is low in inorganic fouling potential which allows the chlorinated AS-RF to serve as a maintenance clean for preventing organic fouling, This kind of maintenance clean is similar to daily CIPs currently being used on other processes.
- The use of PACl as a coagulant did not cause any detrimental effects on the membranes.
- Coagulant doses were controlled to maintain settled water turbidities just below 1.5 NTU. Additional PACl improved settled water turbidities but did not affect membrane filtration performance or fouling rates.

Summary

- Complete pretreatment (coagulation, flocculation and sedimentation) was effective in controlling the membrane fouling rate as the raw water turbidity increased steadily during the operational period. The clarified water turbidity was consistent at 1.5 NTU, at an average coagulant dose of 5.5 mg/L.
- The fouling rate increased drastically when no pretreatment was used. The specific flux dropped from 4 gfd/psi to 1 gfd/psi in 7 days of operation when no pretreatment was provided. The specific flux decline rate was 0.43 gfd/psi/day when the plant was operated without pretreatment compared to 0.2 gfd/psi/day when complete pretreatment was used before membrane filtration.
- Increasing the coagulant (PACl) dose did not reduce membrane fouling rates.

CASE II: ASHLAND MICROFILTRATION WATER TREATMENT PLANT

Plant Description

The Ashland Microfiltration Water Treatment Plant (Ashland) is a 1.2 mgd microfiltration plant with USFilter polypropylene membranes (M10C). The raw water is directly coagulated with ACH and passed through a 500 µm strainer before entering the membrane system. Chlorine is used as disinfectant after membrane filtration. Figure B.4 presents a schematic of the Ashland WTP.

The USFilter microfiltration system consists of four skids, and each skid contains 48 individual membrane modules. The flow is equally distributed to the four skids. The systems are operated with a 45 minutes filtration cycle followed by 2.5 minutes backwash. Chemical cleaning, which is applied approximately monthly, employs phosphoric acid at pH of 2, followed by EDTA, sodium hydroxide, and hydrogen peroxide.

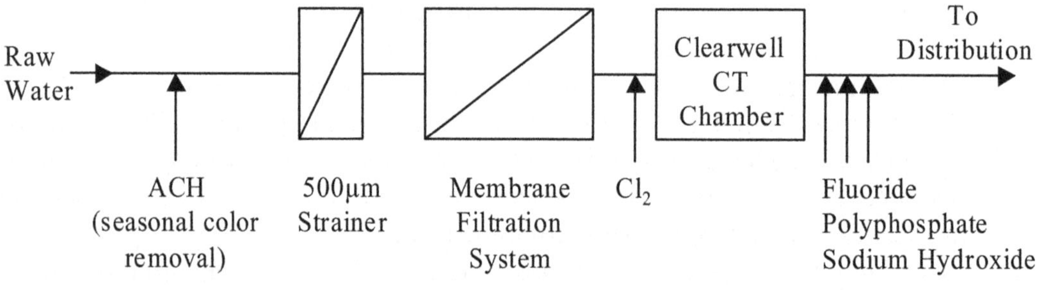

Figure B.4 Ashland Microfiltration Water Treatment Plant schematic

Water Quality

The water quality of this plant is shown in Table B.2. The raw water has low-moderate alkalinity with an average of 48 mg/L as $CaCO_3$. As the raw water is drawn from a lake, the turbidity values fluctuate between 0.5 – 100 NTU, with an average value of 10 NTU. The average TOC is approximately 1.3 mg/L indicating low organic presence in the raw water. Color is a seasonal issue that is mostly made up of colloidal clay solids and dissolved organic compounds.

Table B.2
Ashland WTP water quality

Raw Water	**Values (range)**
Temparature (°C)	10 (0.5-20)
Alakalinity (mg/L as $CaCO_3$)	48
Turbidity (NTU)	10 (0.5-100)
Color (TCU)	5 (0-25)
TOC (mg/L)	1.34
Iron (mg/L)	0.18
Manganese (mg/L)	0.006
Membrane Filtrate	
pH	6
Turbidity (NTU)	0.05 (0.04-0.09)
Color (TCU)	0 (0-5)
TOC (mg/L)	1.2

Effects of Pretreatment

Typically, raw water is pretreated with in-line coagulation using 0.4 mg/L ACH during the months of May and June for color removal. The low ACH dosage was in the range recommended by the membrane manufacturer. Based on the water quality analysis, the color was successfully removed from the maximum value of 25 TCU to 5 TCU with in-line coagulation and membrane filtration. However, the TOC levels between raw water and filtrate remained similar.

The low dose of ACH addition for in-line coagulation not only successfully removed color but also stabilized the membrane filtration. Figures B.5 and B.6 present the TMP data versus time in the same season of 2003 and 2004, respectively. During the membrane operation without ACH addition, membrane fouling became rapid and the TMP increased abruptly after the TMP reached 50% of maximum pressure. However, when ACH was added to the system, the TMP increased steadily and more slowly during the filtration run compared to when ACH was not used. The fact that the trend repeated for two consecutive years further confirmed the conclusion that low dose of ACH could stabilize membrane performance and slow down the TMP increase.

Summary

- In-line coagulation with ACH successfully removed color.
- The low dose of coagulant assisted in stabilizing the TMP and reducing the membrane fouling rate.

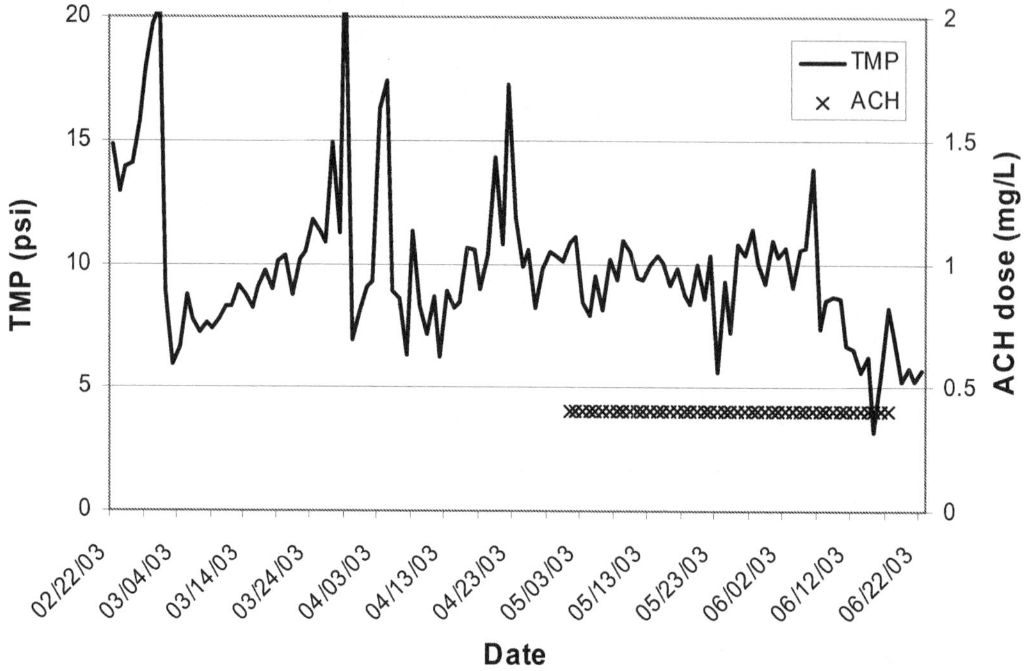

Figure B.5 Ashland membrane TMP data and coagulant dose in 2003

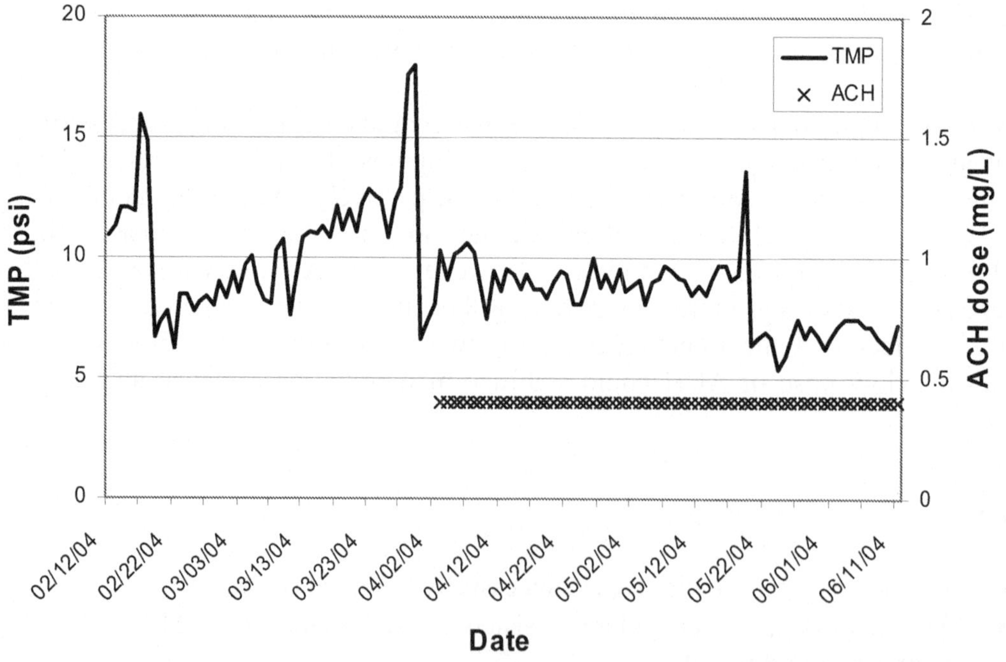

Figure B.6 Ashland membrane TMP data and coagulant dose in 2004

CASE III: MEREDITH WATER TREATMENT PLANT

Plant Description

The Meredith Water Treatment Plant (Meredith) is a 0.6 mgd (2.5 ML/day) microfiltration plant with USFilter polypropylene membranes (72M10C). Raw water is drawn from a 25 ML raw water basin via a 250 mm inlet main. ACH and acid are dosed directly into the inlet main and passed through an in-line mixer, which provides two to three minutes of contact time before the membrane system. Aqueous ammonia is dosed into the filtrate line prior to entering a contact tank. The water from the contact tank is chlorinated and enters the 0.5 ML clear water storage tank before being released on demand to the distribution. Figure B.7 presents a schematic of the Meredith water treatment plant.

The microfiltration system consists of 72 membrane modules. The system is backwashed every 20 minutes, and the backwash waste is transported by gravity from the backwash tank to one of the two 14m x 24m sludge lagoons. Clear supernatant is drawn from the top of the lagoon and returned to the storage basin when the ACH floc settles. The CIP is conducted approximately every two weeks or when the TMP increases to the upper limit, 150 kPa (22 psi).

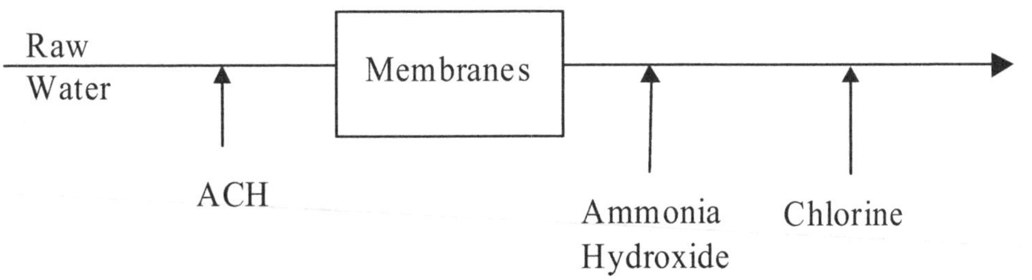

Figure B.7 Meredith Water Treatment Plant schematic

Water Quality

The water quality of the Meredith Water Treatment Plant is shown in Table B.3. The raw water turbidity and color values fluctuated seasonally within a range of 1 – 14 NTU and 5-55 TCU, respectively.

Raw water is pretreated with coagulant (ACH) and an in-line mixer is used to provide mixing of the coagulant. The treatment process, direct coagulation plus microfiltration, successfully removed turbidity from 1-14 NTU to less than 0.1 NTU, color from 5-55 TCU to 5 TCU, iron from 2.5 mg/L to less than 0.3 mg/L, and manganese from 1 mg/L to less than 0.1 mg/L.

Table B.3
Meredith water quality

Raw Water	Values
pH	6.5-8.9
Turbidity (NTU)	1-14
Color (TCU)	5-55
Iron (mg/L)	2.5
Manganese (mg/L)	1.0

(continued)

Table B.3 (Continued)

Membrane Filtrate	Values
pH	7.5
Turbidity (NTU)	<0.1
Color (TCU)	5-6
Iron (mg/L)	<0.3
Manganese (mg/L)	<0.1

Effects of Pretreatment Chemical

Alum was initially used as coagulant at Meredith. However, after a series of testing, ACH was found as a better coagulant. Figure B.8 shows the membrane TMP versus time with alum and ACH as coagulant. ACH reduced the fouling rate by more than 50% compared to the fouling rate when alum was used. With alum, the membrane TMP increased by 60 kPa (8.7 psi) in five days, while the TMP increased 60 kPa (8.7 psi) in more than 10 days when ACH was used.

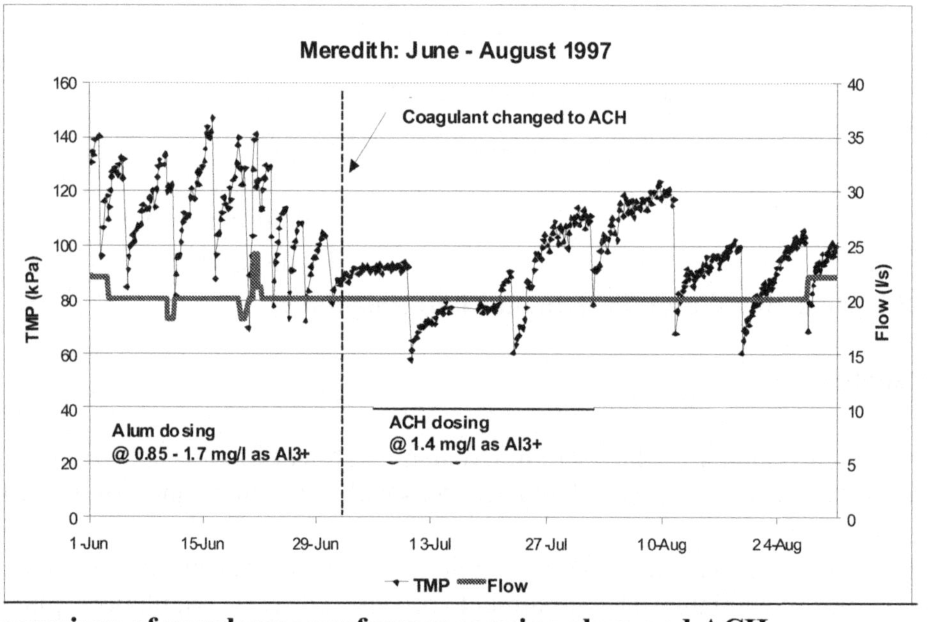

Figure B.8 Comparison of membrane performance using alum and ACH

Summary

- Using microfiltration, with in-line coagulation, was effective in controlling turbidity, color, iron, and manganese.
- When ACH was used as coagulant, the TMP increased approximately 50% less than that when alum was used as coagulant.

CASE IV: WAIKATO WATER TREATMENT PLANT

Plant Description

Waikato Water Treatment Plant has a submerged PVdF membrane system with a design capacity of 12 mgd. Pretreatment to the membrane consists of sludge blanket clarification using alum as coagulant. The plant treats river water and experiences significant spikes in turbidity during storm events. Figure B.9 presents a schematic of Waikato plant. Since the river water has low alkalinity (20 – 45 mg/L as $CaCO_3$), lime and carbon dioxide are added to increase alkalinity to improve coagulation and adjust pH. GAC is used for final polishing of membrane filtrate.

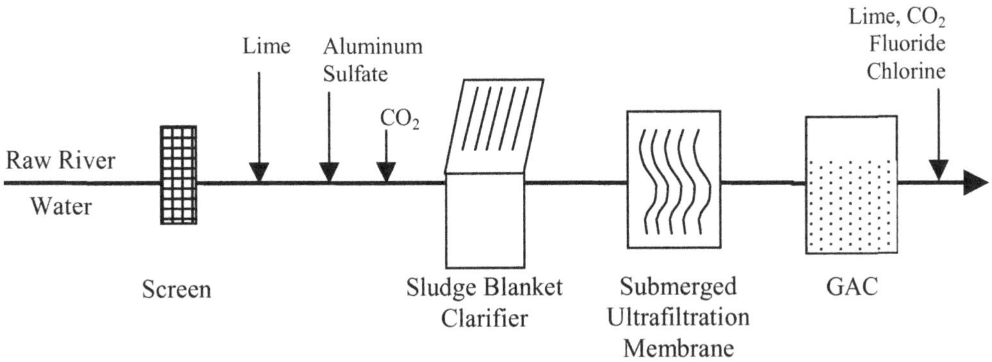

Figure B.9 Waikato Water Treatment Plant schematic

The plant was initially configured with 6 trains of 9 cassettes during startup in July 2002. The number of cassettes per train were subsequently increased to 10 in July 2003 and 11 in November 2003. Recently, two additional trains have been added, for a total of 8 trains, increasing hydraulic capacity to 20 mgd. The increased number of cassettes per train provide greater ability to deal with increased fluxes experienced by trains remaining in operation when a train is taken offline for direct integrity testing, maintenance cleans, or full chemical cleans. The membrane performance goal is to maintain an interval of a minimum of 30 days between full chemical cleanings.

Water Quality

The water quality at this plant is shown in Table B.4. The raw water has low alkalinity, ranging from 20 to 45 mg/L as $CaCO_3$. The turbidity is high with average value of 18.5 NTU and spikes reaching 129 NTU. Total organic carbon (TOC) is moderately high, averaging 4.6 mg/L. The plant experiences seasonal algal blooms with algae counts ranging from $8x10^5$ to $4x10^7$ cells/L. Silica concentrations in the raw water are very high at 28 mg/L on average

Table B.4
Waikato water treatment plant water quality

	Units	Count	Avg	Median	Range
Raw Water					
Temperature	degC	360	16.9	17.4	10.5 - 24.2
pH		360	7.4	7.4	6.9 - 8.4
Alkalinity	mg/L (CaCO3)	42	36	36	20 - 45
TDS (est from Conductivity)	mg/L	360	9.6	9.2	6.7 - 12.6
Turbidity	NTU	360	18.5	15.2	7.2 - 129.2
TOC	mg/L	40	4.6	4.5	2.3 - 8.0
DOC	mg/L	41	4.0	3.9	1.4 - 6.3
UV	1/cm	41	0.098	0.087	0.008 - 0.197
Silica	mg/L	22	28.0	28.6	23.5 - 32.0
Iron	mg/L		0.93		
Manganese	mg/L		0.064		
HPC	cfu/mL	13	6,977	5,250	420 - 32,000
Algae	cells/L	23	8E+06	1E+07	8E+05 - 4E+07
Clarified Water					
Turbidity	NTU	360	1.5	1.3	0.4 - 5.2
Membrane Filtrate					
TOC	mg/L	41	2.4	2.3	1.2 - 4.1
UV	1/cm	41	0.030	0.028	0.013 - 0.065
Turbidity	NTU	360	0.014	0.013	0.000 - 0.030

Plant Performance

Pretreatment Performance

The objective of pretreatment at this plant is the reduction of particulate and organic foulants. The sludge blanket clarifier has been operated reliably and consistently at removing these foulants with average turbidity decreasing from 18.5 NTU in the raw river water to 1.5 NTU in the clarifier effluent, while DOC is reduced from 4 mg/L on average in the raw water to 2.4 mg/L in the membrane filtrate. The sludge blanket clarifier has also been effective at removing algae, another potential membrane foulant. Average pretreatment chemical doses are 2 mg/L lime, 40 mg/L alum, and 15 mg/L carbon dioxide. Pretreatment conditions at this plant have not been significantly changed since plant startup in July 2002. Figure B.10 presents the average daily clarified water turbidity and daily alum dose. The figure shows clarifier effluent turbidity ranging from 0.5 NTU to 2.5 NTU, until late in 2003. Alum dose was in the range of 30 to 40 mg/L with seasonal variations.

While pretreatment has been successful at removing particulate and organic foulants, it has also been determined to play a role in the inorganic fouling of the membranes. Silica present in the raw water at a high level, along with aluminum from the coagulant, combine to form kaolinite. Kaolinite, $[Al_2Si_2O_5(OH)_4]$, was determined through membrane autopsy to be the major foulant.

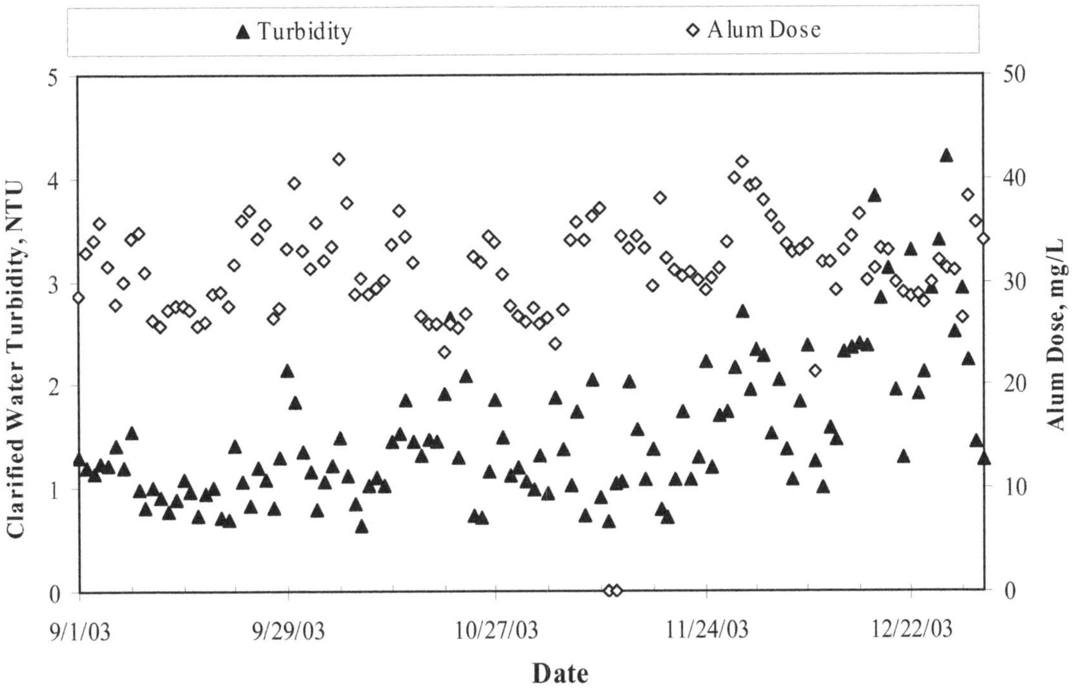

Figure B.10 Clarified water turbidity and alum dose

Membrane Performance

While the plant has consistently met its membrane performance objectives, including 30 days between cleanings, it became apparent after approximately one year of operation that kaolinite fouling increased the CIP frequency. To address this, citric acid maintenance cleans were initiated on an as needed basis to extend the run times. On days that citric acid maintenance cleans were performed, daily sodium hypochlorite maintenance cleans (250 mg/L) were not performed. The citric acid concentration used for maintenance cleaning was 2 g/L (0.2%), producing a cleaning solution pH of approximately 3.

Table B.5 presents the operating parameters and available water quality data for two different filtration periods: high fouling and low fouling. Daily sodium hypochlorite maintenance cleans were performed during the high fouling period, and citric acid maintenance cleans were performed in place of the daily hypochlorite maintenance cleans every fifth day on average during the lower fouling period. The high fouling period occurred in October and November 2002 and the low fouling period was selected from a filtration cycle one year later, in 2003. The important raw water characteristics are similar during both periods.

Clarified water TOC and DOC were not measured in 2002, but DOC averaged 4.5 mg/L during the 2003 filtration cycle. Dissolved silica in the raw water was similar during the two filtration cycles, averaging 25 mg/L during the 2002 and 27 mg/L during 2003. Since the two filtration cycles occurred at approximately the same time of year, other water quality parameters were assumed to be similar.

Table B.5
Operating and water quality data during test periods

Parameter	Unit	High Fouling Period	Low Fouling Period
Pretreatment			
Average Alum Dose	mg/L	26	31
Clarified Turbidity	NTU	1.1	1.2
Membrane Operations			
Average Flux	gpm/sf 20C	44	30
Feed Water Recovery	%	93	93
Raw Water Quality			
Average Temperature	degC	17.7	15.6
Average pH		7.4	7.3
Average UV254 Abs.	1/cm	0.240	0.270
Average Silica.	mg/L	25	27

The effect of citric acid maintenance cleans is evident from Figure B.11 and B.12. Figure B.11 presents hourly specific flux versus time for a filtration cycle in 2002, when no citric acid maintenance cleans were used, while Figure B.12 presents specific flux versus time for a filtration cycle in 2003 when citric acid maintenance cleans were conducted, on average, every 5 days. A representative set of data were isolated for each filtration cycle. These data are plotted separately with open-centered data points.

To account for the fact that the average temperature corrected flux was higher (44 gfd) during the high fouling cycle than during the low fouling cycle (30 gfd), the specific flux data were plotted against volume of water produced per square foot of membrane area. The lower flux operated in 2003 was attributed to the extra cassette added to each train in July 2003, and the different numbers of cassettes used during these two periods are factored into the calculation of the volume of water produced. The resulting chart is presented in Figure B.13. Best fit straight lines were applied to the data in this figure. The slopes of these lines represent the rates of fouling as the loss of specific flux per gallon of water produced, normalized for the membrane surface area and temperature. The fouling rate for the high fouling filtration cycle, when only daily chlorine maintenance cleans were performed (–0.0038) is more than 30 percent higher than the fouling rate when citric acid maintenance cleans were performed on average every fifth day (-0.0029).

The data from citric acid maintenance cleans contained a number of points with specific flux significantly higher than the average value. Often these data points were produced when the hypochlorite maintenance cleans were performed immediately after the day of a citric acid maintenance clean. The hypochlorite maintenance cleans were particularly effective at restoring specific flux at this case.

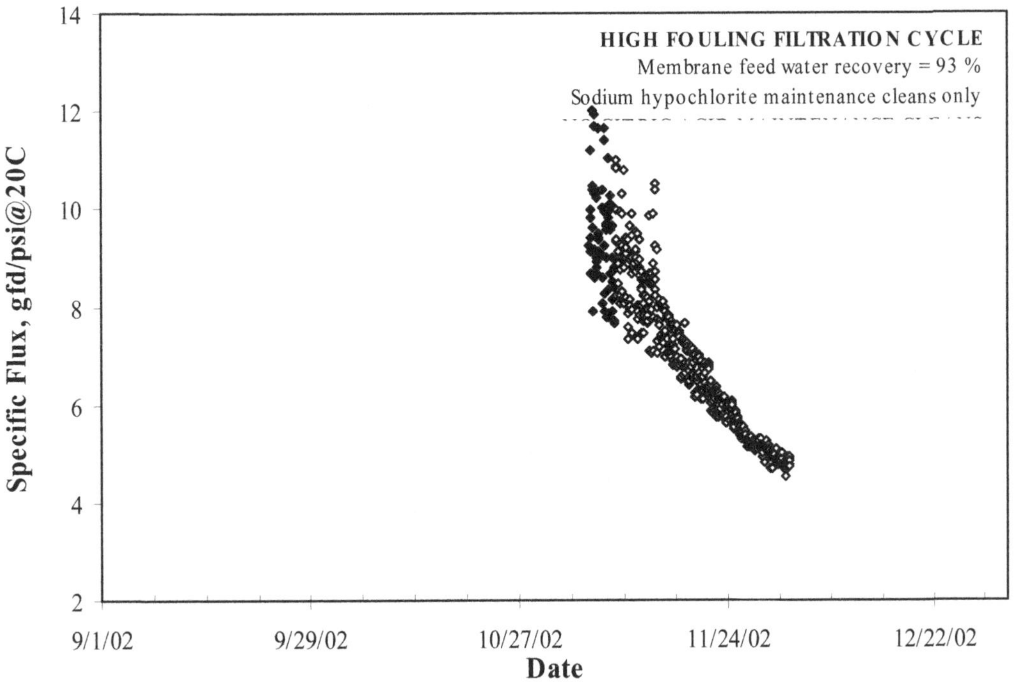

Figure B.11 Membrane specific flux versus time (daily chlorine maintenance cleans only)

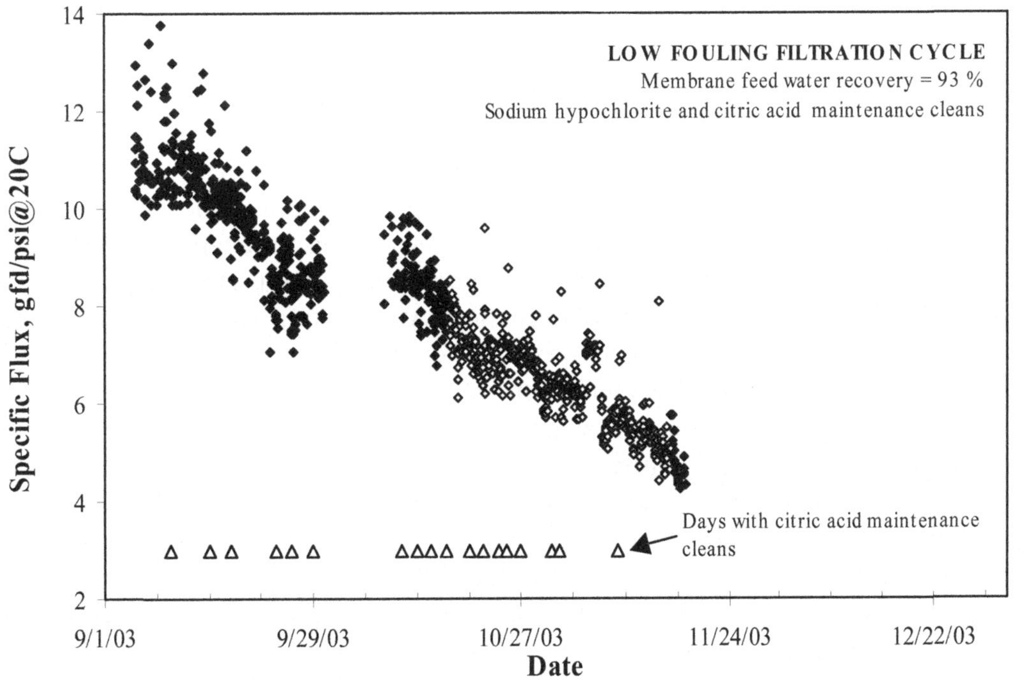

Figure B.12 Membrane specific flux versus time (daily chlorine maintenance cleans with periodic citric acid maintenance cleans)

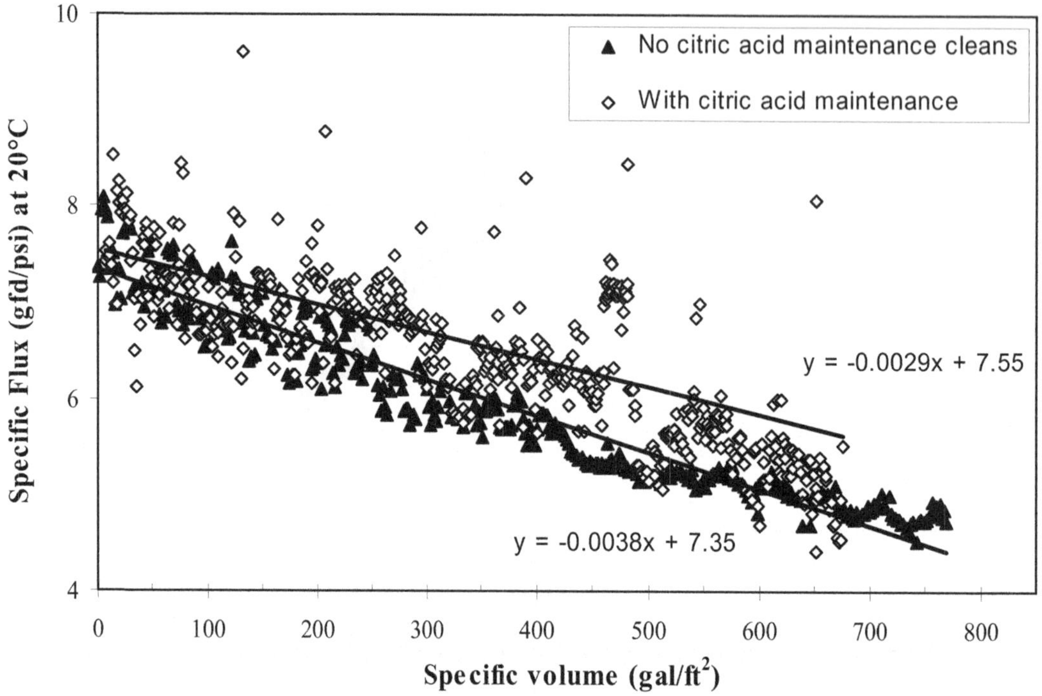

Figure B.13 Membrane specific flux versus gallons water produced per square foot membrane surface

Lessons Learned

- Bench-scale testing of alternative coagulants were performed to establish whether alternative coagulants would be suitable for replacing alum as the primary coagulant, which would prevent the kaolinite fouling. The criteria for changing coagulant were not increasing the operational cost, nor compromising the pretreatment objectives of organic removal to prevent THM formation. After testing with confirmed coagulant dose, it became apparent that it was more economical to undertake more frequent citric acid maintenance cleans and continue using alum rather than make any change.
- The frequency of maintenance cleans changes in line with water demand. At low system demand, there is sufficient redundancy within the membrane system that citric acid maintenance cleans are not required, but a recovery clean is performed every 6 weeks. This has the advantage of reducing operational cost, but also ensuring long term restoration of membrane permeability. As the flow increases to 75% of plant capacity, the citric acid maintenance clean frequency will be 2-3 times a week, and 4-7 times a week when the plant is operated at maximum flow. The frequency of maintenance clean increases towards the end of interval between two recovery cleans, particularly as the membrane permeability drops.
- As the citric acid maintenance clean process was not envisaged at the time of plant design and construction, the resource consent for the waste discharges to the

environment did not identify the cleaning waste. All cleaning chemical solutions, after neutralization, are discharged back to the Waikato River. An Assessment of Environment Effects was undertaken and a change to the consent was granted.

- The biggest issue for the maintenance clean has been optimizing the duration of the citric acid maintenance clean in order to maximize available production time. The in-tank neutralization of the cleaning solution was the biggest challenge and the most time consuming step. As citric maintenance cleans are fully automated, compared to recovery cleans which are controlled by operators, the distribution control system is required to neutralize the liquid waste such that discharge at pH of 6-9 to the Waikato River is permitted. Changes made to the consent was the reduction in pipe runs to instrumentation providing the pH value to the process, the flushing time for dosing pipe work, and the installation of control system alarms that indicate a problem with the process. If the neutralization fails, the train will be taken out of service, increasing flux rates on the remaining trains, negating the benefit of an automatic system.

Summary

- Membrane pretreatment using a sludge blanket clarifier with alum as coagulant and lime and carbon dioxide addition to increase alkalinity and control pH produced a consistently good quality membrane feed water. The turbidity was reduced from 18.5 NTU on average in the raw water to 1.5 NTU in the clarifier effluent, and DOC was reduced from 4.5 mg/L on average in the raw water to 2.4 mg/L in the clarifier effluent.
- Silica present in the raw water at approximately 25 mg/L, combined with aluminum from coagulant addition, to form the membrane foulant, kaolinite.
- Fouling due to kaolinite was controlled by the use of periodic citric acid maintenance cleans. The fouling rate was reduced by 30 percent when citric acid maintenance cleans were performed approximately every 5 days versus a filtration cycle when no citric acid maintenance cleans were performed.
- It was more economical to conduct citric acid maintenance cleans instead of changing the coagulant chemical.

CASE V: BEXARMET WATER TREATMENT PLANT

Plant Background

The BexarMet Water Treatment Plant (BexarMet) near San Antonio, Texas treats water from the Medina River. The 9 mgd (34,000 m^3/day) production capacity plant supplies 125,000 inhabitants of San Antonio. Water production of this facility began in December 1999, and a second phase leading to a double capacity is under consideration.

The treatment line is composed by a settling step with 2 Superpulsators® (sludge blanket clarifier manufactured by Degremont), which is operated with 12% ferric sulfate at a dose of 12 mg/L as weight of commercial product. PAC (less than 10 mg/L) can be added if needed, followed by an Aquasource UF system. A schematic of the treatment train is shown in Figure B.14. Figure B.15 shows the picture of the Superpulsators.

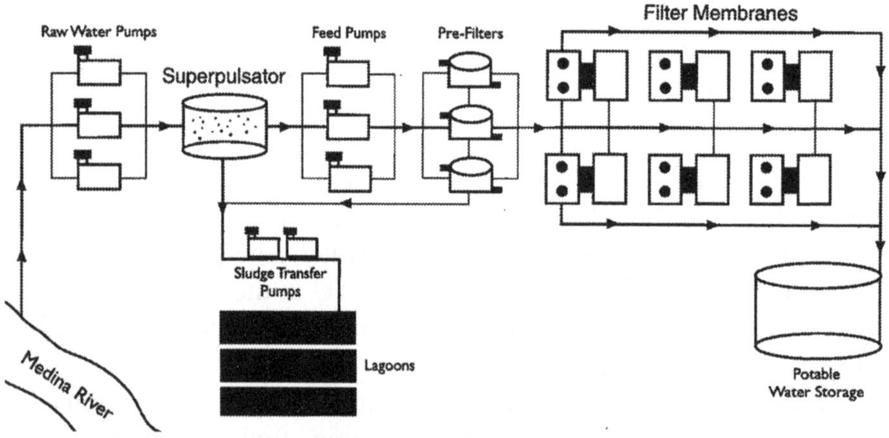

Figure B.14 Schematic of the BexarMet treatment line

Figure B.15 Picture of Superpulsators

The UF backwash water is recycled back to the head of the Superpulsator®, which reduces water loss to less than 1%. The integrity of the UF modules is operated semi-automatically by air pressure hold test, which is monitored by the programmable logic controller (PLC). The pressure decay test is checked weekly or after CIP is performed. The Bexar plant was the first plant which used Aquasource membranes installed immediately downstream after a primary clarification without an intermediate filtration step for surface water. Figure B.16 shows the membrane plant and membrane module. The systems specifications are listed below:

- 3 pre-filters (200 μm)
- 7 UF racks (skids)
- 336 modules total, and each module is 300 mm in diameter with 55 m^2 filtration area
- Total membrane surface area: 18,480 m^2 (199,000 ft^2)
- Design flux: 115 L/h.m^2 (68 gfd) at 20°C (dead end operation mode)
- Backwash interval: 45-90 min
- Recycling of UF backwash water ahead of Superpulsator® clarifiers

Figure B.16 Pictures of membrane plant and membrane module

Water Quality

Table B.6 shows the raw water quality of the Medina River. The relevant characteristics of the raw water can be summarized as follows:

- Highly variable turbidity, with peak values exceeding 6,000 NTU
- Average hardness is 250 mg/L as $CaCO_3$
- TOC concentration ranging from 2 to 11 mg/L
- UV absorbance ranging from 0.02 to 0.98 cm^{-1}

Table B.6
Medina River Raw water quality parameters

Raw Water	Average	Range
PH	8	7.1-8.6
Turbidity (NTU)	30	5-6000
Alkalinity (mg/L as $CaCO_3$)	210	180-372
Hardness (mg/L as $CaCO_3$)	250	150-300
Conductivity (μs/cm)	575	NA
DOC (mg/L)	1.8	1.4-4.1
UV (cm^{-1})	0.05	0.02-0.98

NA: not available

The water quality parameters measured at different points of the treatment line are summarized in Table B.7. The average raw water turbidity was 30 NTU, and the coagulant dosage ranged from 10 to 100 mg/L of ferric sulfate as commercial product weight (40 mg/L as average) was used during this time period. The average settled water turbidity was usually below 2 NTU, and it was approximately 1 NTU most of the time. The settled water UV absorbance was 0.02 cm^{-1} on average. The treatment objectives were achieved even though PAC was not added.

Table B.7
Water quality parameters through the treatment line

Raw Water	Raw Water	Settled Water	UF Permeate
pH	8	7.5	NA
Turbidity (NTU)	30	1.5	0.02
Alkalinity (mg/L as $CaCO_3$)	210	180	NA
Hardness (mg/L as $CaCO_3$)	250	250	NA
TOC (mg/L)	2.1	1.1	1.0
Color (°Hazen)	6	2	1
UV (cm^{-1})	0.05	0.023	0.024

NA: not available

Plant Performance

Figure B.17 and B.18 show the membrane performance of rack 4 from March 2003 to March 2004. The net production flux was approximately 120 L/h-m² which allowed the plant to be operated above the design value. The permeability varied between 180 and 250 L/h-m²-bar at 20°C.

The specific flux decline was noticeable at the last three months of the data set, which was mainly due to the seasonal variation of the raw water quality. The winter season normally has the most challenging water condition for membrane filtration: low water temperature, high turbidity, and high organic concentration. The permeability usually recovers once the warm weather begins. The seasonal impact on the specific flux has been observed over the past few years and could be explained by the broad range of water temperature variations from 10°C to 30°. The water has higher viscosity at lower temperature that corresponds to higher permeability when the flux is corrected at 20°C.

During the high turbidity events with more than 6,000 NTU in the raw Medina River water, the coagulant dosage was adjusted up to 200 mg/L and no severe membrane fouling was observed. No detergent cleaning was needed during the first five years of operation. Maintenance cleaning with citric acid is performed every 10 days.

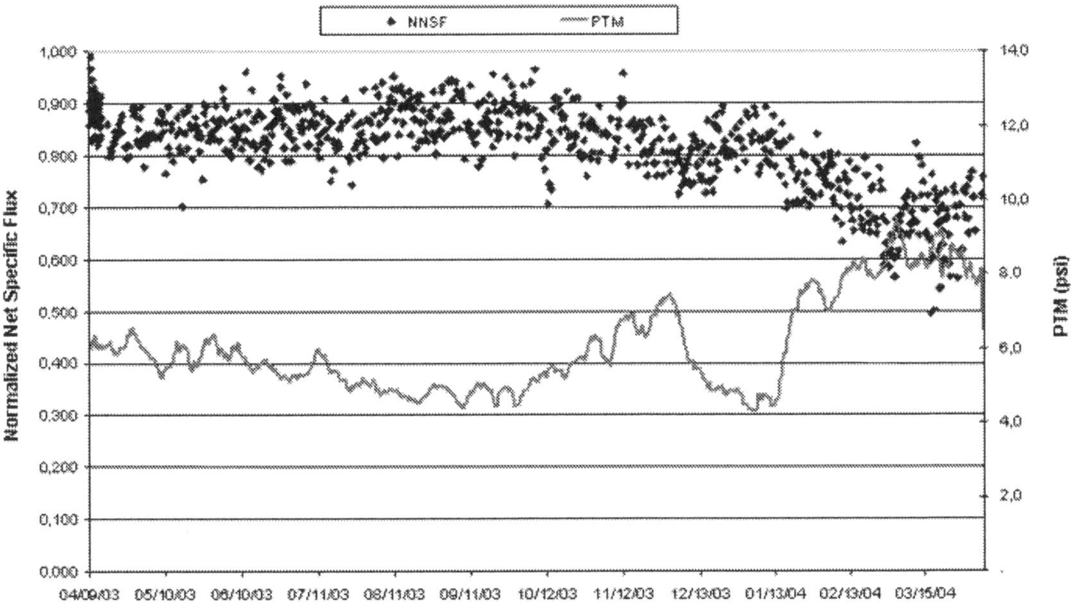

Figure B.17 BexarMet UF plant rack 4 hydraulic performance curve

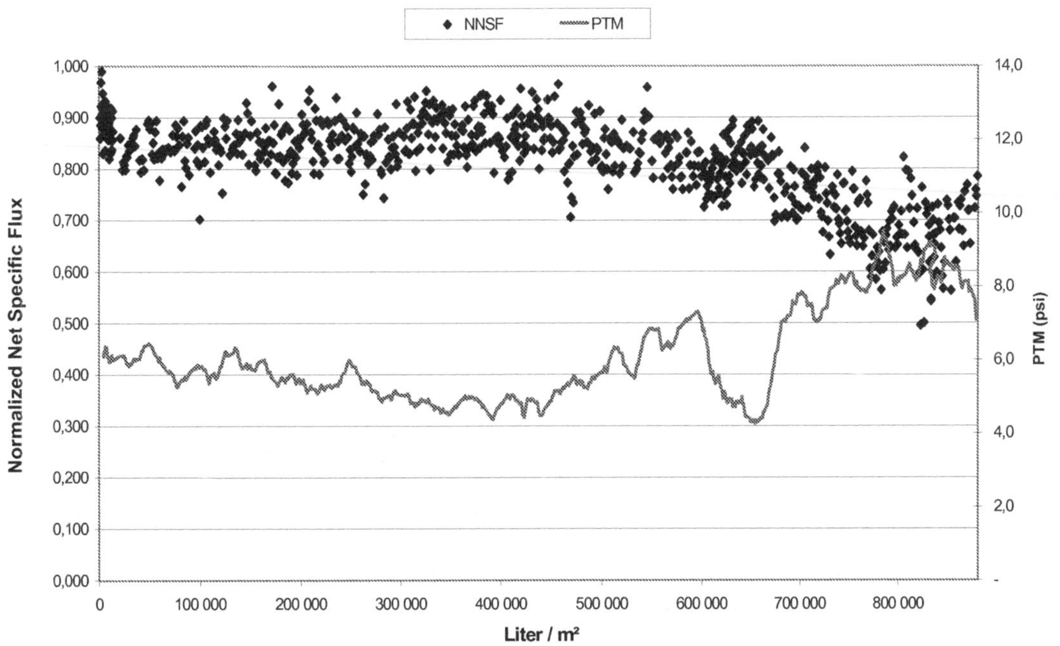

Figure B.18 BexarMet UF plant rack 4 hydraulic performance curve

CASE VI: VIGNEUX WATER TREATMENT PLANT

Plant Background

The Vigneux Water Treatment Plant (Vigneux) is located in France, near Paris, with Seine River water as water source. The plant has been in operation since the end of 1997. Table B.8 shows relevant raw water quality parameters of Vigneux.

This plant was the first to use an innovative hybrid pretreatment process (Cristal process) consisting of the application of PAC upstream of UF membranes and backwash water was recycled back to a floc blanket reactor (FBR). This process was designed to mitigate seasonal events of micro-pollutants (e.g., pesticides, or taste and odors) and to reduce the content of natural organic matter, which was responsible for increasing disinfectant demand and causing disinfection byproducts. The Cristal process was integrated into the existing process consisting of coagulation-sedimentation, granular activated carbon filtration and ozonation. The schematics of the plant is shown in Figure B.19.

Table B.8
Raw water quality parameters of the river Seine at Vigneux WTP

Parameter	*Typical values*	
	Min. - Max.	*Average ± Std*
TOC, mg/L	2 - 10	2.9 ± 0.2
UV	4.5 - 12	5.2 ± 0.6
Turbidity, NTU	3 - 300	6.5 ± 2.7
Temperature, C	2 - 25	14 ± 0.2
pH	7.8 - 8.2	8.0 ± 0.1

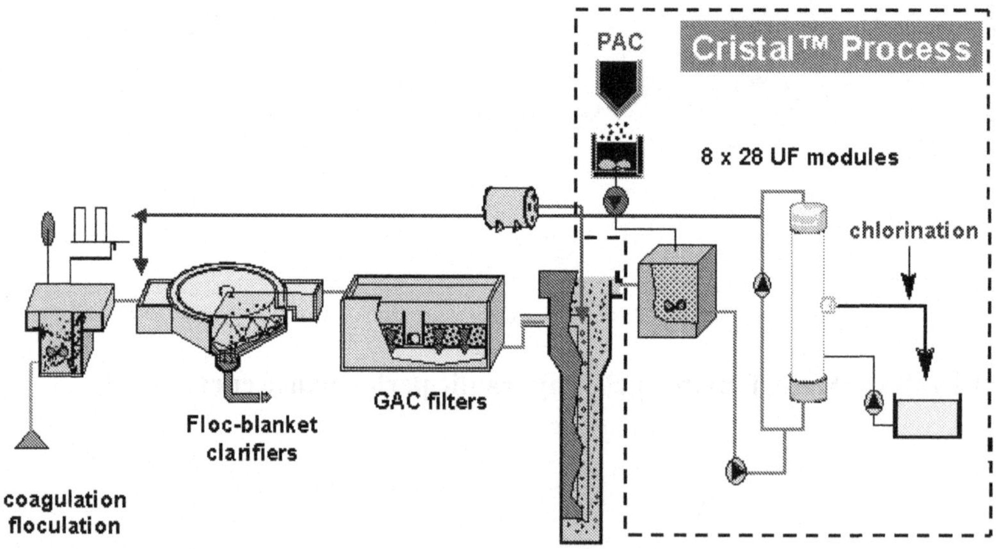

Figure B.19 Schematics of the treatment line at Vigneux plant

The UF system consists of 8 racks each containing 28 modules with 64 m² of filtration surface per module. Figure B.20 shows the inside of the UF plant building. The production capacity of the plant is 14.5 mgd (55,000 m^3/day) at 20°C, which represents a flux of 150 to 180

L/h.m². Although 8 percent of the permeated water is used for backwash, the total water loss of the treatment plant is less than 1 percent due to the recycling of the backwash water to the floc blanket clarifier.

The membrane system was operated in a recirculation mode until May 2004. In order to reduce the operating expenditures (mainly due to energy consumption), the system has been switched to a dead-end mode since May 2004. The average operating flux has been decreased to 130 L/h.m². The current operating conditions of the plant are:

- Filtration time: 60 min
- PAC dosage: 2 – 12 g/m^3 (maximum 7 in dead-end operation)
- Instantaneous flux: 100 - 150 L/h.m^2 (59-89 galday.ft²)
- Backwash chlorination: 5 g/m^3
- Standard backwash with 325 l.h.m^2 (pressure: 1.5 bars minimum and 2.2 bars maximum)
- Coagulant: Aluminum sulfate for the clarification

Figure B.20 View of the UF plant

Plant Performance

Figures B.21 and B.22 show the performance of the rack 3 in the Vigneux plant. The switch of membrane operation from the recirculation mode to the dead-end mode is indicated on the graph. The operating conditions including duration of the filtration cycle, backwash protocol, and PAC dosage were kept the same during the switch, and the specific flux slightly decreased. It can be noticed that the specific flux maintained in a more stable stage after a few months into the new operating mode. Some transient events of rapid decrease of the permeability occurred found to be due to the changes in the raw water quality. The raw water quality changes were caused by variations of the raw water organic content, mainly polysaccharides. The membrane performance was recovered by reducing the water production without the implementation of any chemical cleaning procedures.

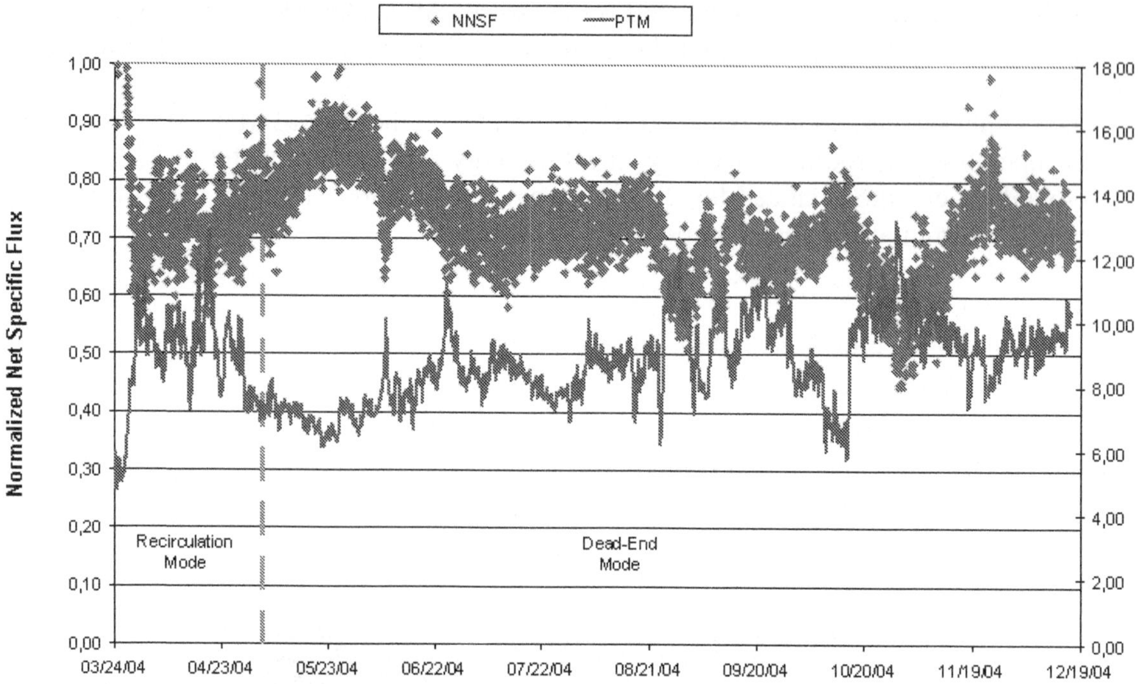

Figure B.21 Hydraulic performance of rack 3 at Vigneux UF plant over the past year 2004 (average filtration flux = 130 l/h.m^2)

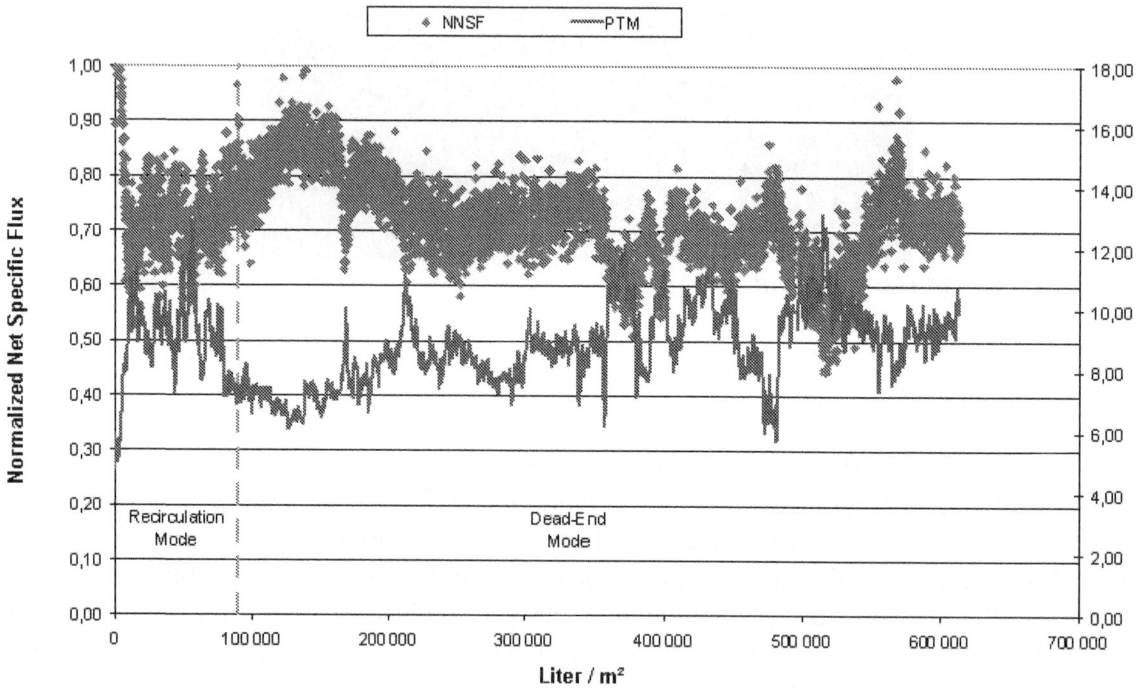

Figure B.22 Hydraulic performance of rack 3 at Vigneux UF plant (average filtration flux = 130 l/h.m^2)

REFERENCES

Adham, S.S., V.L. Snoeyink, M.M. Clark and J.-L. Bersillon. 1991. "Predicting and verifying organics removal by PAC in an ultrafiltration system." *Journal American Water Works Association*, 83(12): 81-91.

Adham, S.S. 1993. Evaluation of the performance of ultrafiltration with powdered activated carbon pretreatment for organics removal. Ph.D. Thesis, University of Illinois: Urbana, IL.

Adham, S.S., V.L. Snoeyink, M.M. Clark, and C. Anselme. 1993. "Predicting and verifying TOC removal by PAC in pilot-scale UF systems." *Journal American Water Works Association*, 85(12): 58-68.

Adham, S.S., J.G. Jacangelo, and J-M. Laîné. 1996. Characteristics and Costs of MF and UF Plants. *Journal AWWA*, 88(5):22-31

Adham, S.S., K. Chiu, K. Gramith, and J. Oppenheimer. 2005. *Development of a Microfiltration and Ultrafiltration Knowledge Base*. Denver, Colo.: AwwaRF and AWWA.

Anselme, C., I. Baudin, and M.-R. Chevalier. 1999. "Drinking Water Production by Ultrafiltration and PAC Adsorption. First Year of Operation for a Large Capacity Plant." AWWA Membrane Technology Conference, Long Beach, California.

Aoustin, E., A.I. Schafer, A.G. Fane, and T.D. Waite. 2001. "Ultrafiltration of Natural Organic Matter." *Separation and Purification Technology*, 22(1-3): 63-78.

Atencio, M., S. Chellam, and S. Adham. 1999. "Treatment of Utah Lake Water with Membrane Processes." AWWA Membrane Technology Conference, Long Beach, California.

Bergman, R.A., R. Porter, and S. Levesque. 2001. "Evaluation of Dual-Membrane Treatment for Large-Scale Water Reclamation at Gwinnett County, Georgia." AWWA Membrane Technology Conference, San Antonio, Texas.

Best, G., D. Mourato, M. Singh, and D. Olson. 2001. "Application of Immersed Ultrafiltration Membranes on High Turbidity and High TOC Surface Waters." AWWA Membrane Technology Conference, San Antonio, Texas.

Bian, R., Y. Watanabe, N. Tambo, and G. Ozawa. 1999. Removal of Humic Substances by UF and NF Membrane Systems. *Wat. Sci. Tech*, 40(9):121-129.

Bourke, M., S. Harrison, B. Long, and T. Lebeau. 2001. "MIEX(tm) Resin Pretreatment Followed by Microfiltration as an Alternative to Nanofiltration for DBP Precursor Removal." AWWA Membrane Technology Conference, San Antonio, Texas.

Braghetta, A., J.G. Jacangelo, J. DeMarco, D. Metz, and J. Swertfeger. 1999. "Maximizing Microfiltration Operational Efficiency Through Transmembrane Flux, Backwash, and Coagulant Addition." AWWA Membrane Technology Conference, Long Beach, California.

Braghetta, A., M. Price, C. Kolkhorst, and J. Thaxton. 2001. "Use of Physical and Chemical Pretreatment Ahead of Ultrafiltration for Surface Water Treatment in San Antonio, Texas." AWWA Membrane Technology Conference, San Antonio, Texas.

Brown, J.C., D.A. Hugaboom, and G.F. Crozes. 2004. "Investigating the Feasibility of a Large-Scale Membrane Filtration Retrofit: A Kansas City, Missouri Case Study." *AWWA Annual Conference*, Orlando, Florida.

Burns, N., M. Horsley, S. Freeman, and T. Schrempp. 2005. "Ultrafiltration Membrane Fouling by Increased Solids and Polymer Carryover from the Pretreatment Process." AWWA Membrane Technology Conference, Phoenix, AZ.

Campos, C., B.J. Mariñas, V.L. Snoeyink, I. Baudin, and J.-M. Laîné. 2000a. "PAC-Membrane filtration process. I: Model development." *Journal of Environmental Engineering*, 126(2): 97-103.

Campos, C., B.J. Mariñas, V.L. Snoeyink, I. Baudin, and J.-M. Laîné. 2000b. "PAC-Membrane filtration process. II: Model application." *Journal of Environmental Engineering*, 126(2): 104-111.

Carroll, T., S. King, S.R. Gray, B.A. Bolto, and N.A. Booker. 2000. "The Fouling of Microfiltration Membranes by NOM after Coagulation Treatment." *Water Research*, 34(11): 2861-2868.

Chang, Y., S. Reiber, and Z. Chowdhury. 2001. "Using Coagulation Assisted Membrane Processes for Arsenic Removal: A Long Term Evaluation." AWWA Membrane Technology Conference, San Antonio, Texas.

Chiu, K.-P., S. Adham, K.J. Howe, C. Mysore, and Z. Do-Quang. 2005. "Bench-, Pilot-, and Full-Scale Studies of Pretreatment for MF/UF Membrane Applications." AWWA Membrane Technology Conference, Phoenix, AZ.

Chiu, K.-P., S.S. Adham, K.J. Howe, C. Mysore, and Z. Do-Quang. 2004a. "Optimization of pretreatment for MF/UF membrane performance." AWWA 2004 Water Quality Technology Conference, San Antonio, TX, AWWA.

Chiu, K.-P., S.S. Adham, K.J. Howe, C. Mysore, and Z. Do-Quang. 2004b. "Optimization of Low Pressure Membrane Pretreatment." *AWWA Annual Conference*, Orlando, Florida.

Clark, S.C., P. Brandhuber, and G. Fuller. 2005. "A Guide for the Selection of Pretreatment When Using Submerged Membrane Systems." AWWA Membrane Technology Conference, Phoenix, AZ.

Cleveland, C., and G. Crozes. 1999. "Solving the Issues of Small System Membrane Treatment Plant Implementation." AWWA Membrane Technology Conference, Long Beach, CA.

Colas, F., V. Bonnelye, and N. Abidine. 2002. "Development of a new hollow-fiber module for large ultrafiltration plants." *Desalination*, 148(1-3): 389-394.

Crozes, G., C. Cleveland, J. Hagstrom, M. Marshall, and G. Rosenbeck. 1999. "Ultra and Microfiltration as Polishing Treatment in Lime Softening Plants Treating Challenging Water Supplies." AWWA Membrane Technology Conference, Long Beach, California.

Crozes, G., and T. Seacord. 1999. "Ultrafiltration Meets Multiple Treatment Objectives in Lime Softening Plants." AWWA Membrane Technology Conference, Long Beach, CA.

Curl, J., and B. Marinas. 2003. "Effect of Operating Conditions on Efficiency of Microbial Contaminant Removal by Low-Pressure Membranes." AWWA Membrane Technology Conference, Atlanta, GA.

Do-Quang, Z., O. Cagnard, M.-R. Chevalier, K.-P. Chiu, and S. Adham. 2004. "Advanced Membrane Operation on Surface Waters: Optimizing Pre-treatment before UF." *AWWA Annual Conference*, Orlando, Florida.

Durand-Bourlier, L., K. Glucina and J.-M. Laîné. 1999. "Assessment of Filtration Mode for Ultrafiltration Membrane Process." AWWA Membrane Technology Conference, Long Beach, CA.

Durand-Bourlier, L., W. Levasseur and J.R. Thaxton. 2001. "From Pilot to Full-Scale Plant Start-Up: The Bexar Facility, Texas." AWWA Membrane Technology Conference, San Antonio, Texas.

Elarde, J.R. and R.A. Bergman. 2001. The Cost of Membrane Filtration for Municipal Water Supplies. In *Proc. of the 2001 AWWA Membrane Technology Conference.* Denver, CO: AWWA.

Elzo, D., I. Huisman, E. Middelink, and V. Gekas. 1998. "Charge effects on inorganic membrane performance in a cross-flow microfiltration process." Colloids and Surfaces A-Physicochemical and Engineering Aspects 138(2-3): 145-159.

Fan, L., J.L. Harris, F.A. Roddick, and N.A. Booker. 2001. "Influence of the characteristics of natural organic matter on the fouling of microfiltration membranes." *Water Research*, 35(18): 4455-4463.

Farahbakhsh, K, C. Svrcek, R.K. Guest, and D.W. Smith. 2004. A review of the impact of chemical pretreatment on low-pressure water treatment systems. *J. Environ. Eng. Sci.*, 3(4):237-253.

Freeman, S. 2001. From Zero to Sixty in Only Seven Years: The Rapid Increase in MF/UF Membrane Surface Water Treatment. In *Proc. of the 2001 AWWA Membrane Technology Conference.* Denver, CO: AWWA.

Fu, L.F., and B.A. Dempsey. 1997. "Effect of charge and coagulant dose on NOM removal and membrane fouling mechanisms." 1997 AWWA Membrane Technology Conference, Denver, CO, AWWA.

Gallagher, P.M., T. Rainier, and S. Lacey. 2005. "Membrane Filtration of Complex Surface Waters: Evaluation of Direct Coagulation and Clarifier Pretreatment Approaches." AWWA Membrane Technology Conference, Phoenix, AZ.

Gallaway, M.A., A.V. Gottberg, and J. Mahoney. 2003. "UF for Sweetwater RO Treatment." AWWA Membrane Technology Conference, Atlanta, Georgia.

Garcia-Aleman, J., K. Mains, and A. Farr. 2005. "Integrating 90 MGD of Membranes, Ozone, and Biological Pretreatment at Lakeview WTP." AWWA Membrane Technology Conference, Phoenix, AZ.

Gumerman, R.C., R.L. Culp, and S.P. Hansen. 1979. *Estimating Water Treatment Costs. Volume 2: Cost Curve Applicable to 1 to 200 mgd Treatment Plants.* Report EPA-600/2-79-162b. Cincinnati, OH: USEPA.

Hackman, B.L., M.M. Clark, K. Carns, and S. Tieman. 1999. "Membrane Technologies for Nitrate and Atrazine Removal from Surface Water Source." AWWA Membrane Technology Conference, Long Beach, California.

Hagmeyer, G., R. Gimbel, and W. Dautzenberg. 2003. "Requirements for Flocculation followed by Ultrafiltration for Direct Treatment of Reservoir Water." AWWA Membrane Technology Conference, Atlanta, GA.

Hall, T., D. Shepherd, E. Glennie, and A. Bird. 2003. "Cryptosporidium Management Using Membranes in a UK Regulatory Environment." AWWA Membrane Technology Conference, Atlanta, Georgia.

Habarou, H., G. Makdissy, J-P Croue, G. Amy, H Buisson, and C. Machinal. 2001. Toward an Understanding of NOM Fouling of UF Membranes. In *Proc. of the 2001 AWWA Membrane Technology Conference.* Denver, CO: AWWA.

Hibbs, S.F., and S. Kendrick. 2003. "Enhanced Coagulation and UF Technology Resolve Compliance Issues for Sweetwater, Texas." AWWA Membrane Technology Conference, Atlanta, Georgia.

Hillis, P., A. Rees, and B. Birkenhead. 2003. "Microfiltration of Surface Water with 99.98% Productivity." AWWA Membrane Technology Conference, Atlanta, Georgia.

Howe, K.J., and M.M. Clark. 2002a. Coagulation pretreatment for membrane filtration. Denver, CO, AWWA Research Foundation.

Howe, K.J., and M.M. Clark. 2002b. "Fouling of microfiltration and ultrafiltration membranes by natural waters." *Environmental Science & Technology*, 36(16): 3571-3576.

Howe, K.J., M.M. Clark, and Y. Wang. 2001. "Effect of coagulation on ultrafiltration membrane performance." 2001 AWWA Membrane Technology Conference, San Antonio, TX, AWWA.

Huey, B., J. Heckler, R. Joost, G. Crozes, and T. Gallier. 1999. "Combination of Powdered Activated Carbon and Low Pressure Membrane Filtration: A Process Alternative for SRP Water Treatment." AWWA Membrane Technology Conference, Long Beach, California.

Humbert, H., H. Gallard, V. Jacquemet, and J.-P. Croue. 2005. "Combination of Coagulation and Ion Exchange for NOM Fouling Reduction of a High DOC Surface Water." AWWA Membrane Technology Conference, Phoenix, AZ.

Jacangelo, J.G., S. Chellam, J. Witko, and N. DeGuida. 1999. "Impact of Fouling on Costs of Integrated Double Membrane System for Treatment of Surface Water." AWWA Membrane Technology Conference, Long Beach, California.

Jack, A.M., and M.M. Clark. 1998. "Using PAC-UF to treat a low-quality surface water." *Journal American Water Works Association*, 90(11): 83-95.

Judd, S.J., and P. Hillis. 2001. "Optimisation of combined coagulation and microfiltration for water treatment." *Water Research*, 35(12): 2895-2904.

Kader, S.A. 2005. "Implementation of Coagulation/Microfiltration for Arsenic Removal: A Case Study at the Fallon Paiute Shoshone Tribe Arsenic Removal WTP." AWWA Membrane Technology Conference, Phoenix, AZ.

Karimi, A.A., J.C. Vickers, and R.F. Harasick. 1999. Microfiltration goes Hollywood: the Los Angeles Experience. *Journal AWWA*, 91(6):90-103.

Kelly, R.F., F. Colas, V. Bonnelye, and S. Tarallo. 2003. "Impact of ultrafiltration membrane material and pre-coagulation strategy on an enhanced clarification process." 2003 AWWA Membrane Technology Conference, Atlanta, GA, AWWA.

Kweon, J., and D.F. Lawler. 2001. "Evaluating softening as a pretreatment for ultrafiltration." 2001 AWWA Membrane Technology Conference, San Antonio, TX, AWWA.

Lahoussine-Turcaud, V., M.R. Wiesner, and J.-Y. Bottero. 1990a. "Fouling in tangential-flow ultrafiltration. The effect of colloid size and coagulation pretreatment." *Journal of Membrane Science*, 52(2): 173-190.

Lahoussine-Turcaud, V., M.R. Wiesner, J.Y. Bottero, and J. Mallevialle. 1990b. "Coagulation pretreatment for ultrafiltration of a surface water." *Journal American Water Works Association*, 82(12): 76-81.

Laîné, J.-M., J.P. Hagstrom, M.M. Clark, and J. Mallevialle. 1989. "Effects of Ultrafiltration Membrane Composition." *Journal American Water Works Association*, 81(11): 61-67.

Laîné, J.-M., M.M. Clark, and J. Mallevialle. 1990. "Ultrafiltration of lake water: Effect of pretreatment on the partitioning of organics, THMFP, and flux." *Journal American Water Works Association*, 82(12): 82-87.

Lee, H., G. Amy, J.W. Cho, Y.M. Yoon, S.H. Moon, and I.S. Kim. 2001. "Cleaning strategies for flux recovery of an ultrafiltration membrane fouled by natural organic matter." *Water Research*, 35(14): 3301-3308.

Lee, J.D., S.H. Lee, M.H. Jo, P.K. Park, C.H. Lee, and J.W. Kwak. 2000. "Effect of coagulation conditions on membrane filtration characteristics in coagulation-microfiltration process for water treatment." *Environmental Science & Technology*, 34(17): 3780-3788.

Lee, N., G. Makdissy, G. Amy, J-P Croue, H. Buisson, and J-C Schrotter. 2003. Morphological Analysis of Low-Pressure Membranes (MF/UF) Fouled by Natural Organic Matter (NOM). AWWA Membrane Technology Conference, Atlanta, Georgia

Levesque, S., D. Williams, and S. Yari. 2003. "Protecting Water Quality in a Georgia Lake Using a Membrane Bioreactor Process with Alum Addition." AWWA Membrane Technology Conference, Atlanta, Georgia.

Lewis, D., B. Fuerst, B. Yallaly, M. Fleming, R. Foster, D. Nix, and R. Bergman. 2001. "Dual-Membrane Treatment for Alternative Water Supplies at Two Texas Municipalities." AWWA Membrane Technology Conference, San Antonio, Texas.

Lin, C.-F., Y.-J. Huang, and O.J. Hao. 1999. "Ultrafiltration processes for removing humic substances: Effect of molecular weight fractions and PAC treatment." *Water Research*, 33(5): 1252-1264.

Lin, C.-F., S.-H. Liu, and O.J. Hao. 2001. "Effect of functional groups of humic substances on UF performance." *Water Research*, 35(10): 2395-2402.

Liu, C., S. Caothien, M. Fushijima, L. Benjamin, T. Otoyo, M.A. Waer, and G. Witcher. 2005. "Combat Membrane Fouling: A Case Study at a 20 MGD Microfiltration Plant." AWWA Membrane Technology Conference, Phoenix, AZ.

Maartens, A., P. Swart, and E.P. Jacobs. 1999. "Feed-water pretreatment: methods to reduce membrane fouling by natural organic matter." *Journal of Membrane Science*, 163(1): 51-62.

Marwah, A., K.J. Howe, K.-P. Chiu, and S.S. Adham. 2004. "Bench-scale studies of pretreatment for membrane filtration." AWWA 2004 Annual Conference, Orlando, FL, AWWA.

McAleese, P., C. Moody, S. Duranceau, B. Corbett, and K. Thorner. 1999. "Integrated Membrane Systems Pilot Test Program for the Olivenhain Water Storage Project." AWWA Membrane Technology Conference, Long Beach, California.

McClelland, S., D. Thomson, E. Jabari, and D. Rohe. 2005. "Chlorine Dioxide and Submerged MF/UF Membranes are In for Surface Waters with High Bromide and Reactive TOC." AWWA Membrane Technology Conference, Phoenix, AZ.

Miller, J., D. Winslet, R.K. Noack, and J. Christensen. 2005. "A New Approach to Low Pressure Membrane Filtration - Direct Coagulation Reduces Costs." AWWA Membrane Technology Conference, Phoenix, AZ.

Mourato, D., G. Best, M. Singh, and S. Basu. 1999. "Reduction of Disinfection Byproducts, Dissolved Organic Carbon and Color Using Immersed Ultrafiltration Membranes." AWWA Membrane Technology Conference, Long Beach, California.

MWH (revised by J.C. Crittenden, R. R. Trussell, D.W. Hand, K.J. Howe, and G. Tchobanoglous). 2005. *Water Treatment: Principles and Design*. 2nd ed. New York: John Wiley and Sons.

Mysore, C.S., J. Leparc, Y. Poussade, J.G. Jacangelo, and D.E. Huffman. 2003. "A Pilot Evaluation of Low-Pressure Membranes for the Removal of Bacteria and Viruses with Municipal Waters." AWWA Membrane Technology Conference, Atlanta, Georgia.

Neemann, J., S. Freeman, and M. Horsley. 2001. "Submerged Membrane Filtration of Coagulated Surface Waters." AWWA Membrane Technology Conference, San Antonio, Texas.

Norton, M.V., T. Richman, G. Fuller, J. Koch, and B. Shaha. 2004. ""No Concrete" Option Doubles the Kennewick Water Filtration Plant's Capacity - the Submerged Membrane Retrofit Approach." *AWWA Annual Conference*, Orlando, Florida.

Pappas, K., J. Pressman, and Z. Chowdhury. 2003. "Use of In-Line Coagulation with Microfiltration at Canyon Lake, Texas." AWWA Membrane Technology Conference, Atlanta, Georgia.

Paxman, S., D. Wetstein, L. Boulos, and S. Adham. 2001. "Evaluation of MF and UF with and without Pretreatment on Utah Surface Water Supply." AWWA Membrane Technology Conference, San Antonio, Texas.

Pickering, K.D., and M.R. Wiesner. 1993. Cost Model for Low-Pressure Membrane Filtration. *ASCE – Journal of Environmental Engineering,* 119(5):772-797.

Richards, A., W. Surrat, H. Winters, and D. Kree. 2001. "Solving Membrane Fouling for the Boca Raton 40-mgd Membrane Water Treatment Plant: The Interaction of Humic Acids, pH and Antiscalants with Membrane Surfaces." AWWA Membrane Technology Conference, San Antonio, Texas.

Roach, D., and J. Vickers. 2001. "Treatment of Nueces River Water Using Conventional Pretreatment and Microfiltration." AWWA Membrane Technology Conference, San Antonio, Texas.

Robert, C., J.S. Taylor, C.R. Reiss, and N. Langenderfer. 1999. "Surface water treatment by integrated membrane systems with in-line coagulation." AWWA Membrane Technology Conference, Long Beach, CA, AWWA.

Roquebert, V., G. Crozes, B. Farver, and S. Luzie. 2001. "Alternative MF/UF Membranes with Chemical Aid for TOC Removal." AWWA Membrane Technology Conference, San Antonio, Texas.

Sahely, B., G. Scott, and R. Filipov. 2004. "Tecumseh Water Treatment Plant Upgrades - Design Issues in the Conversion of a Conventional Filtration Plant to a Membrane Filtration Plant." *AWWA Annual Conference*, Orlando, Florida.

Schäfer, A.I., A.G. Fane, and T.D. Waite. 2001. "Cost factors and chemical pretreatment effects in the membrane filtration of waters containing natural organic matter." *Water Research*, 35(6): 1509-1517.

Schneider, C., P. Johns, and R.P. Huehmer. 2001. "Removal of Manganese by Microfiltration in a Water Treatment Plant." AWWA Membrane Technology Conference, San Antonio, Texas.

Sethi, S., G. Crozes, D. Hugaboom, B. Mi, J. Curl, and B. Marinas. 2003. "Results and Assesment from Full-Scale Testing of Integrity Monitoring Methods for MF and UF Processes." AWWA Membrane Technology Conference, Atlanta, Georgia.

Sharp, M.M., and I.C. Escobar. 2005. "Determining the Effects of Coagulant Pretreatment on Ultrafiltration." AWWA Membrane Technology Conference, Phoenix, AZ.

Shaw, K., J. Lozier, D. Gattis, D. Lewis, and B. Fuerst. 2003. "Dual-Membrane Treatment for Brackish Water Supply at Sherman, Texas." AWWA Membrane Technology Conference, Atlanta, GA.

Shorney, H.L., W.A. Vernon, J. Clune, and R.G. Bond. 2001. "Performance of MF/UF Membranes With In-LIne Ferric-Salt Coagulation for the Removal of Arsenic From a Southwest Surface Water." AWWA Membrane Technology Conference, San Antonio, Texas.

Shrive, C.A., J. DeMarco, D.H. Metz, A. Braghetta, and J.G. Jacangelo. 1999. "Assessment of Microfiltration for Integration into a Granular Activated Carbon Facility." 1999 AWWA Membrane Technology Conference, Long Beach, CA, AWWA.

Singh, M., S. Kendrick, G.J. Best, C. Palnchaud, S.L. Crawford, and D. Bach. 2003. "Simultaneous Oxidation: Coagulation Using Immersed Membranes for the Removal of Dissolved Organics and Iron and Manganese." AWWA Membrane Technology Conference, Atlanta, Georgia.

Taylor, S., J. Henke, S. Jones, and T. Lebeau. 2003. "Evaluation and Selection of a Membrane Water Treatment System to Develop a New Surface Water Supply for Witchita Falls, Texas." AWWA Membrane Technology Conference, Atlanta, GA.

Voges, L.E., M.M. Benjamin, and Y.J. Chang. 2001. "Use of iron oxides to enhance metal removal in crossflow microfiltration." *Journal of Environmental Engineering*, 127(5): 411-419.

White, M.C., L.C. Schideman, M.J. Kosterman, and L.M. Rago. 2004. "Cost-Effective Integration of Membranes into Large Water Plants: A 60-mgd Submerged Membrane Retrofit at the Racine Water Utility." *AWWA Annual Conference*, Orlando, Florida.

Yuan, W., and A.L. Zydney. 1999. "Humic acid fouling during microfiltration." *Journal of Membrane Science*, 157(1): 1-12.

Yuan, W., and A.L. Zydney. 2000. "Humic acid fouling during ultrafiltration." *Environmental Science & Technology*, 34(23): 5043-5050.

Zhang, M.M., C. Li, M.M. Benjamin, and Y.J. Chang. 2003. "Fouling and natural organic matter removal in adsorben/membrane systems for drinking water treatment." *Environmental Science & Technology*, 37(8): 1663-1669.

ABBREVIATIONS

ACH	aluminum chlorohydrate
AFCWTP	Atlanta Fulton County Water Treatment Plant
AwwaRF	Awwa Research Foundation
ATR	attenuated total reflectance
BAC	biologically active carbon
°C	degrees Celsius
CA	cellulose acetate
CCI	construction cost index
CIP	cleaning in place
CIRSEE	Centre International de Recherche sur l'Eau et l' Environnement
CL	cathodoluminescence
cm^2	square centimeter
da	dalton
DBP	disinfection byproducts
DI	deionized
DOC	dissolved organic carbon
EDS	energy dispersive spectroscopy
EDTA	ethylenediaminetetraacetic acid
ENR	Engineering News Record
ft	foot
ft^2	square foot
ft^3	cubic foot
FRP	fiberglass reinforced polyexter
FTIR	Fourier transform infrared
FWR	feed water recovery
g	gram
GAC	granular activated carbon
gal	gallon
gfd	gallons per day per square foot membrane surface area
gpm	gallons per minute
HAA	haloacetic acid
HIOPS	heated iron oxide particles
HPC	heterotrophic plate count
Hz	hertz
kDa	kilodalton

kPa	kilopascals
L	liter
L/day	liter per day
L/mg-m	liter per mg per m
LPDE	low-density polyethylene
m	meter
m^3/day	cubic meter per day
MF	microfiltration
mg	milligram
mgd	million gallons per day
mg/L	milligram per liter
MIEX	magnetic ion exchange process
min	minute
mL	milliliter
mm	millimeter
MWCO	molecular weight cutoff
μm	micrometer
NCCWC	North Clackamas County Water Commission
NF	nanofiltration
NOM	natural organic matter
NSF	NSF International
NTU	nephelometric turbidity unit
O&M	operation & maintenance
PA	polyamide
PAC	powdered active carbon
PACl	polyaluminum chloride
PES	polyethersulfone
PP	polypropylene
PS	polysulfone
psi	pound per square inch
PVC	polyvinyl chloride
PVdF	polyvinylidene fluoride
RO	reverse osmosis
SEM	scanning electron microscopy
SOC	synthetic organic chemical
SUVA	specific ultraviolet absorbance
TCU	true color unit
THM	trihalomethane

TMP	transmembrane pressure
TOC	total organic carbon
TTHM	total trihalomethane
UF	ultrafiltration
UNM	University of New Mexico
USA	United States of America
USEPA	United States Environmental Protection Agency
UV_{254}	ultraviolet absorbance at a wavelength of 254 nanometers

IWA Publishing's authorised EU representative for General Product Safety Regulations is Diane D'Arras, 15 rue Duret, 75116 Paris, France, e-mail: safety@iwap.co.uk.

Printed and bound by CPI Group (UK) Ltd, Croydon, CR0 4YY
17/09/2026
02218798-0001